# We can solve climate change with you!

Each of us can make an important contribution. That's the promise of *Solve Climate Change Now: Do What You Love for a Healthy Planet*.

So, what can you do? Consider the following:

1. **Read and share the message**. People need to know that there's hope and things they can do.
2. **Write a review** on Amazon and share your feedback with reading groups and others. This encourages prospective readers.
3. **Put the ideas into practice**. There are simple things you can do to get started. For example, in less than 10 minutes you can have a Climate Action Conversation with someone to help them find the sweet spot between what they love to do and what's needed for climate health. There's no pushing or cajoling. It's pure attraction.
4. **Identify groups** that might be interested in the message. I offer online and in-person keynotes, play spaces, and other inviting opportunities for business, community, and faith-based groups. www.SolveClimateChangeNow.com
5. **Other ideas to enjoy together?** I'm all ears. Climate@DonMaruska.com

*Thanks for caring and making a difference for generations to come.*

# SOLVE CLIMATE CHANGE NOW

## *Do What You Love for a Healthy Planet*

**Don Maruska**
MBA, JD, Master Certified Coach

**Solve Climate Change Now: Do What You Love for a Healthy Planet**
www.SolveClimateChangeNow.com
Copyright © 2022 Don Maruska

Paperback ISBN: 979-8-817053-03-6

All rights reserved. No portion of this book may be reproduced mechanically, electronically, or by any other means, including photocopying, without permission of the publisher or author except in the case of brief quotations embodied in critical articles and reviews. It is illegal to copy this book, post it to a website, or distribute it by any other means without permission from the publisher or author.

References to internet websites (URLs) were accurate at the time of writing. Authors and the publishers are not responsible for URLs that may have expired or changed since the manuscript was prepared.

Limits of Liability and Disclaimer of Warranty
The author and publisher shall not be liable for your misuse of the enclosed material. This book is strictly for informational and educational purposes only.

Warning – Disclaimer
The purpose of this book is to educate and entertain. The author and/or publisher do not guarantee that anyone following these techniques, suggestions, tips, ideas, or strategies will become successful. The author and/or publisher shall have neither liability nor responsibility to anyone with respect to any loss or damage caused, or alleged to be caused, directly or indirectly by the information contained in this book.

Publisher
10-10-10 Publishing
Markham, ON   Canada

Printed in Canada and the United States of America

# Praise for *Solve Climate Change Now*

"This is a fun and useful book. Research shows when people take one action to fight climate change, they are more likely to take additional actions. *Solve Climate Change Now* has a lot of ideas for getting us all out of our seats to act. I highly recommend this book and Don Maruska's wonderful way of communicating possibilities!"

**~ Jim Thompson, Founder of Positive Coaching Alliance, Co-Founder of Recovery Café San José, and Founder of THIS! Is What We Did**

"Helping people have fun in what they are doing has demonstrated success in energizing the volunteers we need to promote climate health. While some people may fear that climate change is too serious to be playful, it's what we need to remain resilient and sustain this important work."

**~ Tamara Staton, Greater Pacific Northwest Regional Coordinator and Education & Resilience Coordinator for Citizens' Climate Lobby**

"Don Maruska's *Solve Climate Change Now* is exactly what we have been yearning for…an accessible and positive way forward to reduce environmental damage and repair global climate. He outlines meaningful ways in which everyone can engage and make a difference, starting with our own daily activities and then expanding our efforts with teams or groups of similarly concerned people. His unique concept of making this "fun" is pure Don…a reflection of the ways he has helped organizations and individuals develop a mindset that is at once alluring and actionable. Building on his vast coaching and leadership experience, he offers a practical and instructive guide for individual readers, group facilitators, and leaders interested in positive change. This is a book that will both inspire you and lift your spirits."

**~ John Steinhart, Silicon Valley human resources consultant and former Director, Sloan Executive Program, Stanford Graduate School of Business**

"Don Maruska is a dynamo. He inspired our 'Plant It Forward' program and charted a leadership course for our club to become net carbon zero. I hope community service organizations everywhere will read *Solve Climate Change Now* and discover the joy of taking action."

**~ Brian Metcalf, Community Services Chair, Rotary San Luis Obispo de Tolosa**

*Praise for Solve Climate Change Now*

"Unlike most calls to climate action, *Solve Climate Change Now* invites us to find joy as we work together to address climate emergency. This wonderful little book shows that each of us has an important role to play in restoring climate health, and explains how to find delight and meaning in the process. I encourage everyone to read *Solve Climate Change Now* and take action."

**~ The Rev. Dr. Margaret Bullitt-Jonas, co-editor,** *Rooted and Rising: Voices of Courage in a Time of Climate Crisis*

"In order for our planet to thrive, people must thrive too. Learning to align and leverage the hats we wear, the roles we fill and the ways we enjoy spending our time the most with the people whom we care about most with climate action is our best chance to do just that. I'm hopeful that you will take the message of this book to heart and contribute to the awareness, actions, and advocacy we all need to thrive."

**~ Kelly DeMarco, occupational therapist, healthcare and sustainability leader, climate coach, and family farmer**

"Don's research gives light to fun and approachable ways to engage in climate activism."

**~ Calla Rose Ostrander, co-founder of Nerds for Earth and catalyst for creating climate healthy communities**

"Coaching has enabled me to see situations from a fresh perspective, expand my networks, and explore new solutions. I highly recommend finding someone to be your *climate action catalyst*. They help boost your impact, and you have fun doing it."

**~ Nicole Rycroft, founder and Executive Director, Canopy Planet, winner of 2020 Climate Breakthrough Award**

"St. Benedict's Episcopal Church is dedicated to creation care. We hope our example shared in *Solve Climate Change Now* will inspire faith-based communities and secular groups everywhere to embrace their calls and their abilities to heal our planet. As you build awareness, take action, and advocate, you will be serving as valued stewards of creation."

**~ The Rev. Dr. Caroline Hall, Rector, St. Benedict's Episcopal Church, Los Osos, CA**

*Praise for Solve Climate Change Now*

"Personal consumption decisions influence more than 80% of U.S. carbon emissions. Thus, it is essential that we reduce these emissions if we are to meet our climate goals. We can all participate in reducing these emissions. Based on many years of experience as an entrepreneur and a Master Certified Coach, Don Maruska provides a lively, new approach to steps everyone can take towards improving climate health while having fun doing it."

**~ John Horsley Ph.D**

*I dedicate this book to the wonderful gift of nature, which offers us so many blessings by nurturing our lives and healing our planet. I give thanks to the people and organizations who are helping us rediscover and restore our relationships with creation.*

# Table of Contents

Foreword .................................................................................xiii

Chapter 1: We Can Solve Climate Change Now ........................1

**SECTION I: START WITH WHAT YOU LOVE ...........21**

Chapter 2: Shift from Fear to Hope – *The Mental Path* .........25
Chapter 3: Choose Sufficiency as Your Abundance
– *The Freedom Path*................................................................39
Chapter 4: Focus on Actions Rather than Words
– *The Change Path*..................................................................45
Chapter 5: Identify What You Love – *Your Fun Factor*..........53

**SECTION II: CONNECT WITH CLIMATE HEALTH NEEDS ................................................................65**

Chapter 6: Promote Awareness – *Track Your Carbon Footprints*..................................................................................69
Chapter 7: Identify Direct Actions – *Live More Richly* ..........83
Chapter 8: Find Advocacy Opportunities – *Share Your Story* .......................................................................................99
Chapter 9: Select Needs That Interest You – *Apply Your Needs Filter* ........................................................................................109

**SECTION III: ENJOY YOUR CLIMATE SWEET SPOT FOR A HEALTHY PLANET** ......................................117

Chapter 10: Create Your Action Portfolio – *Celebrate and Share Your Success* ................................................................121

Use the Quick Start Guide for Climate Action Conversations ........................................................................133
Notes ...................................................................................143
Acknowledgments ................................................................151
About the Author .................................................................159

# Foreword

The news about the impact of climate change can be terrifying. On the one hand, you may hear people lamenting that there's no hope. On the other hand, you are being urged to take action in bold ways. Trying to figure out where to turn and how to help can seem daunting, even overwhelming.

You aren't alone in your frustration and worry. Millions of other people are frozen in indecision when the earth needs them to be involved: voluntarily, joyfully, and effectively.

Would you like to know how to find enjoyment in taking action to help the climate? Would you like your efforts to be joyful and sustainable, so that you love what you are doing, and also make a difference? Do you want an easy-to-follow guide that starts with you, and what gives you meaning and satisfaction, and then builds from there?

*Solve Climate Change Now* is here to help you reach those goals in an effective, engaging, and enjoyable way. Don Maruska distills decades of scientific insights and proven success as a Master Certified Coach to help you find your fulfilling path to action. His approach will get you thinking and doing in rewarding ways. I can't imagine anyone better than Don Maruska to guide you.

The book shares real stories of real people from different walks of life, and different places across the U.S. and beyond. These will stimulate you to imagine new possibilities, and enjoy this powerful approach to being a climate catalyst.

I hope you will join me in taking action to realize the opportunities that *Solve Climate Change Now* offers you.

**Raymond Aaron**
***New York Times* Bestselling Author**

# Chapter 1

# We Can Solve Climate Change Now

*"The sweet spot is where duty and delight converge."*
~ Thomas Mann

# 1

We can solve climate change now. By "we," I mean you, me, family members, friends, coworkers, community organizations, faith-based groups, and more. We have the capacity. What we need are ways to engage and connect with what's needed and will sustain our efforts. This book offers you practical, proven, and productive ways for us to solve climate change now.

## *If we can solve climate change now, why haven't we yet?*

Part of the problem is that we have the wrong "we." We expect our leaders, scientists, tech entrepreneurs, or someone else to solve it. Wouldn't that be wonderful? If that were the case, we may fancifully think that we wouldn't have to do anything different ourselves. We could simply do what we are doing, and the problem would go away.

But others aren't solving climate change for us. It's not for lack of ideas. Lots of voices are telling us what we "should" do. But we're not responding quickly enough. We're like the frog in a slowly heating pot of water, but the frog doesn't jump out.

Concerned people rail that their leaders don't understand the enormity of climate change or don't follow through on taking the bold actions needed. I'm not surprised that our leaders aren't solving climate change for us. Early in my career, I served as a legislative assistant in the U.S. House of Representatives and the U.S. Senate. Working where the rubber met the road, between legislative interest and results, it became clear that leaders rarely get out ahead of their constituents. If they do, it may be briefly. If their constituents aren't behind them, there'll be no parade, and little will change.

What we lack is enough people with the self-motivation to sustain the awareness, actions, and advocacy needed to gain real solutions. We can't be just "preaching to the choir." We need more "people in the pews." More than that, we need more people taking action. So, how do we crack that challenge?

### *Why is climate change such a tough nut to crack?*

Not surprisingly, many of us confront climate change with dread. We feel the pain of people suffering from the effects of climate change. We worry about our children and their children. Perhaps we worry about wildfires or floods hitting close to home. Or we share concern for those already living in desperate circumstances because of climate change. The challenges of climate change have become so great that some people have become depressed to the point of inaction.

Climate change poses the classic challenge of a chronic problem. We respond much better to acute physical crises—a child in peril, an enemy attacking, someone challenging us. We are wired to solve acute, life-threatening problems. Fear fires up the amygdalae within our brains; signals transmit to release powerful chemicals like adrenaline that give us a burst of energy to fight or flee. We jump into and solve acute problems or suffer immediate consequences. Even a frog dropped into hot water knows to jump out. The issue resolves quickly. But our bodies can't safely sustain such a powerful response for a long time. If we stay in the fear mode, the chemicals released will become toxic to our organs. We will burn out.

In short, an acute, fear-based approach to climate change, which harps on what we "should" do, is not working. At least, it's not working sufficiently to get adequate results. From a marketing perspective, this is a "push" model of change. That is, it's trying to "push" people to take action. However, people don't like to be pushed. Especially in our hyper-politicized and divided communities, people don't want to be told by someone else what they should do. Lamentably, this is even the case when what's to be done may be good for them as well as the planet. We need a new way to solve climate change.

### *How does doing what you love solve climate change?*

The tough nut of climate change also holds the seed of a solution. To find the seed, I harken back to Professor Michael

Ray's Creativity in Business course at Stanford Business School.[1] One of the themes to live with was "do what you love and love what you do." When we do what we love—what we enjoy doing—we have the self-motivation to sustain us. It also means that we have the grit to work through the challenges of our choices. We tap our native creativity and sustain our efforts. This energy, this way of being, is exactly what we need for a challenge as worthy and demanding of our efforts as climate change.

In my work helping people solve tough issues together, I've seen the power of engaging people around their hopes and what gives them joy, rather than dwelling on their fears and avoiding action. Neuroscience supports the idea that a hopeful frame of mind enables us to engage the prefrontal cortex, the executive function of the brain, and the cerebral lobes that enable us to develop creative, new approaches. That positivity doesn't sugarcoat the challenges but instead looks to what we can do and how we can make a difference. We can break away from fearful thinking, where our amygdalae keep us entrenched in fight, flight, or freeze modes. You'll read in Chapter 2 about how you can use this mental path to rewarding results.

Doing what you love is not only about what you enjoy. It's also about what you value and sustain. Chapter 3 focuses on how you can give yourself the freedom to have and enjoy abundance for yourself and others.

This approach to solving climate change embodies the "attraction" model. It appeals to what you love and value. It invites you to be part of the solution. It attracts rather than pushes you to participate. It engages your self-motivation, which is your sustainable energy for action.

Some people may resist giving themselves permission to do what they love and enjoy. They may wonder why we are talking about doing what we love when climate change is such a serious issue. Experience shows that this attraction approach unleashes positive results. In workshops where I ask people about their experiences when they do something they love or enjoy, they light up. They use words like "feel energized," "lighthearted," "blissful," "smile," "laugh," "time flies by," and "flow with ease." People who feel like that are primed to learn and act. It's a vehicle through which they can more readily change their behaviors. And most importantly, enjoyable experiences enable us to do this together with a positive spirit rather than rancor.

For example, consider a reduction in car trips to reduce greenhouse gas emissions. How can that be enjoyable? Some people may fret about how that could be a deprivation. Rather than thinking about what I'm giving up and how I don't want to constrain my freedom, I can think about what would be enjoyable about reducing trips. For example, is there someone that I would like to travel with on one of my trips, so that I would be reducing my trips and increasing the joy of friendship?

Not surprisingly, what's enjoyable for one person may not necessarily be enjoyable for another person. We have different styles, different approaches. So, part of this book is to help you focus on what you love. What do you enjoy? What kinds of things give you joy? And then, how can you align those with what's needed to create climate health?

**Can you really do what you love and contribute to a healthier planet?**

Sometimes it's difficult to feel confident that doing what we love will have the impact that we hope to have in the world. It may seem selfish to focus on that which we love when planetary life is suffering so. But as we'll explore more deeply in later chapters, aligning our actions with our purpose in a way that brings us joy, enables us to be in service longer to the greater good. It avoids burnout. It keeps our inner fire lit. And when our soul is alive, we are best able to offer our gifts to the world in a way that helps us effectively reach our goals.

Consider Tamara Staton's experience. While she often wonders about her efficacy and continually strives for big impact, doing what she loves has sustained her climate work for over ten years. It allows her to balance the moments of burnout and overwhelm with her long-standing commitment to secure a healthy future.

Before entering the climate space, Tamara enjoyed teaching middle and high school and leading wilderness trips with youth because she could be crazy, goofy, and let her wild side run free. She found that playful approaches like storytelling, group games, and interactive projects kept easily distracted young minds engaged and learning. Eager to dedicate more of her life to climate work, she wondered if there was opportunity for engaging and motivating adults through lightness and play in the climate space—a serious topic with committed grown-ups. "Will I be taken seriously or just be seen as a crazy goofball with ridiculous ideas?" she wondered.

Over time, the answer became clear. Instead of seeing playful approaches as conflicting with serious climate action, Tamara found ways to integrate them. She began by starting the Portland Chapter of the Citizens' Climate Lobby (CCL), a national grassroots nonpartisan organization that focuses on national policies to address climate change. In her work over the years as the Greater Pacific Northwest Regional Coordinator, she saw ways to share the support that neuroscience insights offer for the positive benefits of fun in motivating and sustaining action. Skeptics learned that it wasn't Tamara just being goofy. She was tapping the interest and commitment of well-intentioned people who struggle to stay committed over time—including herself. Leaning into this desire, she envisioned and ultimately organized Climate Camp—a two-day workshop designed to integrate joy, leadership, and effective climate action.

Realizing the appetite for resilience-building strategies in the climate space, and how much she would enjoy leaning into this space, Tamara proposed the idea of a resilience coordinator in CCL, and they offered her a position as Education and Resilience Coordinator soon thereafter. As part of the Resilience Building

Action Team, with her leadership support, CCL volunteers developed and expanded an Active Hope program inspired by Joanna Macy's book with that same title. Additional programs, trainings, and resources were expanded to support volunteers online, where Tamara was able to expand her love of teaching, coaching, and facilitating. In one of their CCL Resilience Circles, as an example, she chose to invite participants to complete their version of "Something I love about being alive on Earth today is……" She also invited them to share stories about what got them involved, and used discussion rooms for volunteers to share their experiences about what keeps them motivated.

### Joy Brings Forth Extra Energy

*Tamara (fourth from left, front)
engages diverse team members in playful ways.*

In short, Tamara kept her hope alive to bring what she loves to climate action.

Like Tamara, I've wondered if I could find opportunities to do what I enjoy in the climate action space. My joy lies in conceiving a plan, developing tools to support it, and coaching others to implement it. I can do the implementation work, but I don't especially enjoy it. Rather than try to do everything, especially roles that don't suit me, I've kept to what I love doing and found people who love the complementary roles.

Let's see how this plays out with some additional examples. Calla Rose Ostrander enjoys finding needs and connecting people with them. It's the thread that ties her many community activities together. Brian Metcalf likes being with other people and getting things done. Jim Thompson gets a charge from taking a stand, protesting, and insisting upon change. No one activity is going to attract all three of them. But as we'll see, there is a role for each of them.

## *How do you find climate needs that fit for you?*

While there are many different climate needs to fill, you can think of them in three broad categories: awareness, actions, and advocacy. I call them the Triple A of climate health.

### *AAA of Climate Health*

The triple A of climate health identifies multiple ways to have an impact.

**Awareness:** Learning and sharing information about what drives climate health, our carbon footprints, and potential solutions

**Actions**: Steps you can take at home, work, and in your community

**Advocacy**: Policies you can support for climate health

There's a chapter on each of these opportunity areas (Chapters 6-8). You'll learn about the opportunities and what's involved. This will help you choose the climate needs that appeal to you. Chapter 9 helps you apply a filter to find what you'll enjoy.

### *Where's your "climate sweet spot?"*

When you match what's joyful for you with what's needed for climate health, you'll find your "sweet spot" for making a difference. That's what will sustain you through the many obstacles that chronic challenges like climate change pose. As Frederick Buechner puts it, we are called to the "place where your deep gladness and the world's deep hunger meet."[2]

## Figure 1. Your "Climate Sweet Spot"

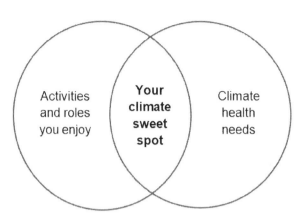

Some people have a proclivity toward one of the A's. Calla Rose targets awareness. She co-founded Nerds for Earth, a volunteer organization that uses technical skills to help rebalance our climate. Calla Rose guides farmers, policy makers, and the public across the U.S. to see how agriculture can shift from a major source of greenhouse gas emissions to a safe and natural way to keep carbon in the ground. Brian Metcalf gravitates toward direct action. He leads tree planting with a local Rotary club and youth groups to reduce greenhouse gases and boost stewardship. He's making a hands-on difference while also broadening the base of people engaged in taking meaningful action. Jim Thompson gives special attention to advocacy. He founded "THIS! Is What We Did," and organizes people to boycott big banks that finance fossil fuels. He puts himself on the line to cut off the flow of capital that feeds greenhouse gas emissions. Each person has found a calling that's joyful to them and makes a difference.

Other people like to play across multiple areas of awareness, actions, and advocacy. Especially effective teams engage members with differing interests to multiply their impact. Importantly, they enhance their impact by attracting and working with others who find their opportunities enjoyable and rewarding.

The steps in this book will help you find your "sweet spot" to solve climate change. If you are already active in advancing climate health, the book will nourish and reinforce your efforts and provide ways to attract others.

## *Why "you?"*

We can only control our own actions. Too many people lay the responsibility for climate action on someone else. We wait for bigger greenhouse gas emitters to act first or for legislators to force them. We imagine that our personal actions won't make a difference or that we have no civic power to influence corporations and politicians. In short, we are trapped in a victim role. Like abuse victims, we have developed learned helplessness. Now is the time for each of us to take charge and do what we can.

But what difference will our individual, household, community, or business actions make? The impacts are two-fold. First, those actions add up. For example, we know that reducing trips with greenhouse gas-emitting vehicles will improve our air quality. If each licensed driver in the United States eliminated one 30-mile trip each month, we would eliminate 40 billion pounds of CO2 each year.[3]

Second, our efforts have ripple effects. *Solve Climate Change Now* invites each of us to look for enjoyable ways to act in our personal, household, business, and community activities. What we initiate and practice in one area will encourage us to take actions in other areas. People who become aware of their greenhouse gas (GHG) emissions at home will become more alert to opportunities at work. And people who see the climate actions we take may be inspired to take similar steps in their own lives. People who act in community groups will call for more action from government and will advocate for the systemic change that is essential to preserve a habitable world.

### *Will this really lead to bigger results?*

Yes, it will. Some people say that this attraction model for climate health sounds utterly naïve. They believe that the enormity of climate change and its myriad threats to our planet require the forceful imposition of Draconian measures. I don't agree. Not because I don't understand the problems. Rather because the push model isn't delivering sufficient results and doesn't fit with human nature.

When we come from a place of hope and engagement with the wonders of creation, we are tapping into the elemental sources of energy that can fuel our individual and collective efforts. Let's explore how this works.

The AAA framework for climate health can get us there. Research shows that people who encourage and model individual changes to reduce carbon emissions gain more support for policy changes.[4] Through behavior, people become engaged. Through engagement, they become advocates.

**Figure 2. AAA Cycle Generates Climate Health**

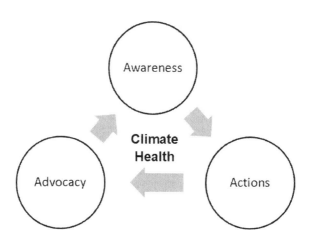

The attraction model works especially well at the community level. It's there that people have pre-existing relationships and can work together to make a difference. Indeed, an international research team identified the community-urban level as the optimal level for impact.[5] Chapter 9 highlights that the tipping point for large-scale action doesn't require a majority of public to engage. An active proportion of 25% or less can bring about major changes. Thus, there is a path for us to follow that leads to the larger systemic changes we need.

As the adage asks, "If not now, when? If not you, who?" Let's boost the attractiveness of solving climate change to the people needed to support results. Let's celebrate the progress of people doing what brings them joy and yields results. When we do, the magnetism of positive change will strengthen. Leaders will join the parade, and we'll all benefit.

### *How can you benefit from this book?*

Newcomers wondering what they can do about climate change, seasoned climate workers, and burned-out activists sitting on the sidelines will all benefit from this book. You'll find it to be a tonic for these troubling times.

As you've probably already gathered, this is not a book that gives you prescribed actions. While the book doesn't offer pat answers, it delivers a process for you to find the answers that are right for you. It also provides concrete examples to prompt your thinking about what's possible.

Throughout the book, there are references to resources at the website, www.SolveClimateChangeNow.com. That's where you'll find free templates for tools like the Fun Finder, Needs Filter, and Opportunity Sorter, and links to information and organizations that can help you on your path. I also encourage you to share what you've learned and would like to offer to other readers.

### Who can guide you?

When you share this journey with family members, friends, coworkers, or community members, you will learn more, move faster, and have more fun. Think of them as catalysts for you. You may remember from chemistry that catalysts stimulate a reaction and accelerate results, but don't become used up in the reaction. In an analogous way, what I'm calling *Climate Action Catalysts* are people who ask the questions posed in this book, reflect back what they hear from you, and help stimulate your actions and thinking. Playing with others in this way will enhance your enjoyment and effectiveness.

The "Opportunities for you to enjoy," at the end of each chapter, invite you to reflect on what you've read. Take some time to respond.

At the end of the book, you'll find a Quick Start Guide for Climate Action Conversations. These conversations pull together what you love to do and climate needs to identify your climate sweet spots. In less than 10 minutes, you can get started on your action path.

### Where can you find support for the journey?

When you work with others in your workplace, service organizations, or community, you'll accelerate and multiply your efforts. When you pursue your climate actions through groups where you are already a member, you have built-in support. Thus, climate actions that you enjoy become another opportunity to accomplish results with people with whom you already have good relationships and can even strengthen those relationships.

There are many opportunities to create climate action teams at work, through social service organizations like Rotary, Kiwanis, and others, and with faith-based groups that share your interest in a healthy planet. Actions by you, your friends, and colleagues will mobilize more people and actions. Together, you will reach the tipping point to solve climate change for yourself and generations to come.

## *Opportunities for you to enjoy*

Suspend judgments about what you have told yourself or what other people have told you about how you "should" act for a healthy climate.

Reflect for a few moments about when you have felt fulfilled in your personal life or work. Did what you were doing feel fun or joyful? How has that made a difference for you?

Identify family members, coworkers, and community organizations with whom you'd enjoy sharing this experience. You don't need everyone to join immediately. Some may be ready now. Others may become interested later. As you attract more people, you'll grow organically and keep thriving with the attraction model for sustained action.

Find one or more persons with whom you can be Climate Action Catalysts. As you read this book and experiment with what it offers, listen and reflect what you hear from one another.

*Enjoy the journey!*

# SECTION I
# START WITH WHAT YOU LOVE

# Start with What You Love

We start with what you love because you are the engine of change. Too many people don't choose to take action for climate health because they don't see themselves doing what a particular action may require. When you discover what's really important to you and gives you hope, you'll have the sustaining energy you need to make a difference. When you do what you love, you'll love what you do. That is, you'll be willing to do what's needed.

Mastery will require building skills in three areas. The three paths are mental, freedom, and change. Think of them as paths to enjoy.

**Figure 3. Finding What You Love**

```
   ┌─────────┐
   │ Mental  │
   │  Path   │
   └─────────┘
        +
   ┌─────────┐         ┌─────────┐
   │ Freedom │    →    │  What   │
   │  Path   │         │   you   │
   └─────────┘         │  love   │
                       └─────────┘
        +
   ┌─────────┐
   │ Change  │
   │  Path   │
   └─────────┘
```

# Chapter 2

# Shift from Fear to Hope – *The Mental Path*

*"We have nothing to fear but fear itself."*
~ Franklin Delano Roosevelt

More than 40% of Americans felt "disgusted" or "helpless" about climate change, according to a 2020 survey published by researchers at Yale University.[1] How can we expect to solve climate change when so many people have essentially dropped out? This challenge requires us to take a fresh look at how to find solutions.

Rarely does anyone ask us what we love doing. In our culture, we tend to ask people what they do for work. We rarely explore what they enjoy about the work that they do—the parts of their job that make their work feel like play. Indeed, if we keep thinking a healthy planet is something we must "work on," it will keep us from finding the joy we need to power ourselves forward.

**Let go of your attachment to the problem and move to solutions.**

The challenge we face with climate change is our fixation on the problems rather than on solutions. Certainly, understanding the problems is an important element of finding solutions. However, when we focus so incessantly on the problems, we deprive ourselves of the very energy that we need to move forward with practical steps.

As we'll see in Section II of this book, there are many resources that offer solutions. Indeed, there are available solutions that could resolve our climate issues and yield climate health. It's clear that we need to fret less and do more.

## *Get your "buts" out of the way.*

When I ask people about climate change and what they might do to promote climate health, I hear a lot of "buts." They sound something like this: "I could be doing, or I'd like to be doing, something for climate health, but I don't know what would be helpful." Or, "What difference can I make?" "Other people or institutions is where the action lies. So, go get someone else to do this work." In short, "It's not my job, and it doesn't sound like fun."

Which of the following "buts" stymie you?

"I'd like to improve climate health, **but** ...

- ... I don't know what to do.
- ... I don't have time.
- ... Our group faces special obstacles that are especially difficult to overcome.

- Others are causing more of the problem. Start with them first.
- It's really someone else's job to do (government, etc.).
- It's all so overwhelming.
- I'm not an expert, and I don't have any influence.
- It's too late for action, anyway. It feels hopeless."
- [Add your own.]

Do you notice how in each of these statements the person has given up their power? They are fearful that they can't do anything or won't be successful. While each of the statements may have some truth to it, this mode of thinking isn't likely to galvanize people into action. Rather, it is likely to feed their amygdalae and lead to fight, flight, or freeze responses. In short, we have a rampant mental problem getting in the way of taking fulfilling actions.

What if we took a lesson from the great improvisors of the world and replaced the "buts" with "ands" and followed a playful approach?

"I'd like to improve climate health, ~~but~~ **and** ...

- I don't know what to do.
- I don't have time.
- Our situation is special.
- Others are causing more of the problem. Start with them first.
- It's really someone else's job to do (government, etc.).

... It's all so overwhelming.
... It feels hopeless."
... [Add your own.]

Now, shift your attitude and take a joyful approach to exploring the challenge.

"How interesting! What can I enjoy doing with this situation?"

What can I learn about the obstacle? Is it real? What are possible ways to overcome it? Could I simply step around the obstacle and swing into action? This shift in thinking will release a whole cascade of ideas and opportunities.

Let's examine a "but" to taking action to reduce our personal carbon footprints. The most common argument is that focusing on our own carbon footprint lets the big greenhouse gas emitters off the hook. As a result, people don't take the small steps they can in their own lives. But regardless of the much greater emissions from the fossil fuel industry, we know that when we take positive action where we can, we're not excusing them. Instead, we're demonstrating our refusal to be constrained by the constructs that others create. We build a foundation of personal integrity that empowers us. We can say that we've done what we can. We can more effectively demand that businesses, legislators, and others take responsibility for what they can do.

For example, employees who cut their personal emissions have a stronger position from which to call upon their employers to do likewise. It's akin to corporations matching their employees' charitable donations.

Facing climate changes of unprecedented dimensions in modern human history, we need to improvise our way forward. Try playing a game of Climate Improv to loosen your grip on the "buts" that are holding back you, your family, and friends from the powerful actions you can proudly take.

**Connect with your hopes—what's really important to you.**

Where can we get the mental energy, the will, to joyfully pursue the solutions before us?

If you want to enjoy what you are doing, you need hope. At least, if you want sustainable enjoyment, you need to have something that powers your mind and actions in a positive direction. And if your efforts are going to come from your own motivation, then they need to come from your hopes.

Hope has gotten a bad rap in some circles. It sounds to some people like wishful thinking or a fairy tale world. True hopes are hopes that arise from deep within us about what's important to us at a core level, versus those that are wishful pursuits of what we think would be nice to have. Making contact with our deep hope stimulates clearer thinking. We've seen it work with tens of thousands of people around the world. When we ask them what their hopes are about a topic, and why those hopes are important to them, they tap their inner motivation.[2] Often, taking the opportunity to delve deeper and deeper stimulates creative thinking and collaborative action.

First, it's critically important to distinguish between expectations and hopes. Expectations carry a fixed attachment to a particular way of doing things or a particular outcome. Hopes are deeper and more fundamental.

How can you tap the mental power and sustaining motivation that hope offers? It can be straightforward. You can invite someone to ask you a series of questions and reflect back what you say. This stimulates your thinking and enables you to find your true interests, what will give you joy, and what will be your fun.

Here's a sample dialogue between Beth, who is concerned about her child's health, and a friend who is serving as her Climate Action Catalyst (CAC).

CAC: What's a hope you have about climate health?

Beth: I want my daughter to have the kind of stable, invigorating climate that I enjoyed growing up.

CAC: Why is that important to you?

Beth: I want her to be healthy and able to thrive.

CAC: Why is that important to you?

Beth: A healthy climate will help her be healthy, and that's the most important gift I can give her.

CAC: OK. So, you'd like your child to grow up healthy and strong, and a healthy climate is important for that.

Beth: Exactly.

Do you notice that Beth started off with an expectation for her daughter, a fixed attachment to the climate Beth had growing up? Since we know that even under the best of circumstances, the climate changes or varies, being tied to how things were decades ago is not realistic. But having a healthy climate so that her child grows up healthy and strong is a clear hope. It's Beth's aspiration for her daughter.

Helping each person connect with their self-motivation creates the foundation for sustainable results. You will be helping them engage their productive thinking.

## *Declare freedom from fear*

Climate health is important, not only for the planet but also for our mental well-being. Recent studies show that fear of the impact of climate change has dramatically diminished public health.[3]

Even if our initial personal actions don't have immediate, dramatic effects, they set in motion shifts in attitude. As we see some progress, we become encouraged to take more actions. The diagram below illustrates the dynamic. Our thoughts frame our actions that, in turn, influence our results. And the cycle continues so that we develop new habits of thinking that empower us to act and make a bigger difference.

**Figure 4. Shift Your Mental Path**

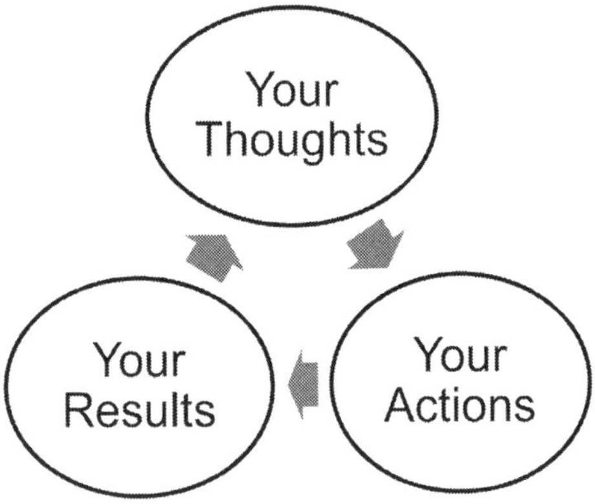

```
        Your
      Thoughts
      ↙      ↘
  Your    ←   Your
 Results      Actions
```

Take some time with people who can support you to clarify your hopes and why they are important to you. Observe how the discussion shifts the dynamics.

## *Connect, collaborate, and create community for fun*

You can gain traction on the mental path with the help of other people. As they say, if we're feeling uneasy or depressed about something, the best thing to do is to go help somebody else. The same is the case here. If you gather with others and focus on sharing your hopes for a healthy climate, you'll find shared energy and motivation to go forward.

For example, in working with members of our Rotary club, seeking to take climate action, I invited them at the start of the planning meeting to go around and have each person share a hope they have for a healthy climate. They gave voice to the following:

- Better understand the impacts of climate change and where and how we can make a difference.
- Achieve measurable progress toward climate health with hands-on projects.
- Provide practical and relevant individual and club opportunities that fit with members' differing living and work situations.
- Enhance quality of life (for example, better transportation options to reduce traffic, etc.).
- Integrate kids into projects with educational and stewardship components.
- Celebrate and share our results.

This brief conversation not only engaged each person with something personally motivating but also helped to define desirable activities. All of that flowed from a simple question and from generously listening to one another.

You can do the same with the people and organizations available to you. If you are uncertain, try it as an experiment. See what a simple question can do to reframe your thinking and that of your friends and colleagues. Observe how this improves your sense of optimism, creativity, and motivation to act.

## *Opportunities for you to enjoy the mental path*

- Practice changing the "buts" you face, to "ands," and focus on the opportunities they offer. Play Climate Improv to free up your thinking. Be fascinated about the obstacles. What can you do with them?

- Identify your hopes for a healthy climate. Have other people to draw out why those hopes are important to you. Discover the depths of sustainable self-motivation.

- Declare your freedom from fear. It's not that climate change isn't scary. Instead, it's that you can choose to place your intention on what you can do.

- Find a group of people at work, in your community, or beyond, who share your hopes for joyful participation in actions that promote a healthy climate.

Congratulations! You've made progress to master *the mental path* needed to solve climate change now.

# Chapter 3

# Choose Sufficiency as Your Abundance – *The Freedom Path*

*"Sufficiency isn't two steps up from poverty or one step short of abundance. It isn't a measure of barely enough or more than enough. Sufficiency isn't an amount at all. It is an experience, a context we generate, a declaration, a knowing that there is enough, and that we are enough."*
~ Lynne Twist, *The Soul of Money*

## Do the new world math

Did you know that if everyone on the planet were to live the American lifestyle, we'd need four to five planets worth of resources?[1] In addition, we'd need all the energy required to produce, distribute, use, and, ultimately, dispose of those acquisitions. Imagine the greenhouse gas emissions to fulfill the aspirations of people around the world to acquire the coveted "First World" lifestyles. We have indulged lifestyles that the planet can't afford for others to adopt.

**Figure 5. Unsustainable Consumption**

We'd need 5 Earths of resources for everyone on the planet to have the average American's consumption.[1]

Enlightened leaders and organizations understand that we are on an unsustainable and unfair path. I enjoy working with such people because they get it. It's in their value system to think of equity, fairness, and sharing. They have done the work with the mental path described in Chapter 2. They realize that accumulation is simply unbridled fear that there won't be enough, and leads to greed to acquire more. They have connected with their hopes to break the cycle of fear.

## *Make less your more*

> "Live simply so that others can simply live."
> ~ Mahatma Gandhi

Have you had an experience of less being more? Maybe you've been on an open beach or under the stars in a wilderness. Did you feel that you were lacking something? Or did you glory in the abundance of nature before you?

Consider the following choices to live simply yet abundantly. What would you like to put on your list of climate friendly ways that feel abundant for you?

**Figure 6. Add Your Ways to Live Simply Yet Abundantly**

| Climate Unfriendly | Climate Friendly |
|---|---|
| GHG vehicles (e.g., fossil fuel RVs, ATVs) | Tents, hiking, backpacking |
| Large amounts of red meat | Sustainable fish, white meat, local greens |
| Single use plastic containers | Reusable glass containers |
|  | Your choices: |

## Choose Sufficiency as Your Abundance

### *Slim down what you're carrying and have a better journey*

As you shift from your fears to your hopes, what can you let go of in your life? What's no longer meaningful to you? What have you accumulated that takes more to manage and attend to than the joy it gives?

As our family approached these questions, we stripped our home of energy-sucking gadgets we don't use or don't really need. Out went the extra power strips to keep heaters, chargers, etc. going when not needed.

More deeply, we became more mindful of our choices. We consolidated laundry loads to reduce energy use. We enjoyed more greens and ate red meat sparingly.

### *Define your new world abundance*

If all of this sounds more like deprivation than fun, please revisit your deepest hopes that you identified in Chapter 2. Instead of thinking about whether you are willing to let go of something, you could create a zero-based budget for a lifestyle aligned with what you need in order to fulfill your hopes. By that, I mean demonstrate to yourself that you have what you need to realize your deepest hopes.

Here's an example:

> *"I find nature to be a perfect home. The sky above is the roof, the air is fresh and provides enough ventilation, and the plants and animals serve as the best form of entertainment."*
> ~ Michael Bassey Johnson, *Song of a Nature Lover*

## *Opportunities for you to enjoy the freedom path*

- Reflect on a time when you felt free. What was the experience like? What flexibility and options did it open for you?
- Experiment with the zero-based, hopeful budget. What are the basics you'd need to enjoy a fulfilling life?
- What would you like to do to live more simply, so that others may simply live? What support would help you to keep moving in that direction?
- How are your reflections helping you identify what you truly love?

Enjoy the freedom you give yourself.

# Chapter 4

# Focus on Actions Rather than Words – *The Change Path*

*"Start with changing behaviors, not mindsets.
It is much easier to 'act your way into new thinking'
than to 'think your way into new actions.'"*
~ Jon Katzenbach

In decades of helping people find ways to work together and resolve tough issues, I've learned the value of simply getting people together to do things. Good processes for interacting and taking action outperform philosophical arguments.

## *Change behaviors; it's easier than changing minds*

As the old saying goes, "The only person who wants to be changed by someone else is a wet baby." This is especially true about changing people's minds. We see this play out in the divisive "culture wars" occurring on many topics—climate, race, education, etc.

It goes back to the dynamics of fear and conflict discussed in Chapter 2. When we challenge people on their beliefs, we threaten their self-esteem.

Instead, people might try a new behavior, especially if it doesn't require a big ongoing commitment to get started. For example, even technophobic grandparents who wouldn't go to a software training class, will try smartphones and apps to connect with their grandkids.

### Let people choose their own rationale

Interestingly, even some climate change deniers are willing to plant trees. Some might enjoy the beauty of the trees. Being in the outdoors may attract others. Additional interest can come from wanting to do activities with family and friends, or gaining recognition. The key is not requiring everyone to think, believe, and espouse the same rationale. We multiply the opportunities for working together when we let people find their own way to the action.

Some people might fear the implications of a line of thinking. "Where will this lead?" "What might I have to give up?" "What could be taken away from me?" Do you hear the fear-based thinking shutting down the prospects for productive action?

Action is the antidote to angst. Going light on rhetoric and heavy on action can cut through the ego conflicts of who's got the right idea and who's thinking correctly or incorrectly about something.

### Get skin in the game to deepen caring

Action also builds caring, and caring builds commitment. When we invest interest, time, and energy, we naturally care more. We also open ourselves to learning more. For example, three years ago, our Rotary club planted a grove of 80 live oak seedlings to honor a dedicated member who died. It was one tree from each member of the club. The effort included a commitment to water and tend the trees during the critical first three years of their growth.

At the time of the planting, the emphasis was on honoring the fallen member and providing a community service. During the three years, the members learned that at maturity the oaks will sequester an estimated 260 tons of $CO_2$. This prompted them to think about other people to honor by planting trees. As a takeoff from Catherine Ryan Hyde's best-selling book *Pay It Forward*,[1] I suggested that we expand the tree planting with a "Plant It Forward" program. Since the trees we plant literally benefit generations to come, we are truly planting it forward. Plus, with an interest of each member in honoring their children, grandchildren, and other loved ones, there's a strong motivation for everyone to participate.

Maybe such tree planting could accumulate sufficient living carbon in their roots to offset some or all of the planters' household greenhouse gas emissions. Imagine leaving their descendants with a clean personal carbon ledger!

Brian Metcalf embraced this opportunity. His story underscores the power of action. In attracting Rotary club members to plant more trees as part of the Rotary Climate Action Team, Brian provided easy opportunities for people to spend a few hours together. He emphasized camaraderie and the joy of accomplishing something together. In short, he went light on proselytizing and heavy on doing. Brian kept people informed about the progress, sharing photos of the seedlings and arrival of the grow tubes to support them. People enjoyed becoming part of the ongoing story that their actions helped write. They had fun doing projects with Brian.

### "Plant It Forward" in Action

Rotarians planting seedlings to benefit generations to come

*Focus on Actions Rather than Words*

## Be conversational and invite people to try something with you

It doesn't take a lot of words or explanation to get started. You can simply invite someone to look at something with you— maybe an action you've started. Use an easy and effective conversational form to invite their participation.[2]

For example, I wanted to encourage others to join me in reducing energy use. So, I expressed an intention that created a bridge with others on our Earthcare Team at church. "Hey, I'd like to share with you something I'm enjoying doing to cut my energy use and save money. I think you'd find it interesting." Then I shared some observations. "I found that by simply accumulating our laundry and reducing wash and dry loads by two each week, I have saved over 400 kilowatt hours and $100 a year. Plus, we don't have the sound of the machines running as much." The approach concluded with an invitation or request. "Would you like to look at some actions you can try in your home?"

You can ease into action in inviting ways.

## *Opportunities for you to enjoy the change path*

- List some approaches that encourage you to take action.
- Identify some experiences you've had where drawing people together to act brought about positive change.
- Outline what makes a project "change friendly" and joyful for you and others.

Observe how action energizes yourself and others.

## Chapter 5

## Identify What You Love – *Your Fun Factor*

*"Creativity is intelligence having fun.
Having fun is the best way to learn."*
~ Albert Einstein

While doing what you truly love involves things that are of value to you and of sustaining interest, fun is an important component.

## *Tap your fun factor*

Fun is not only enjoyable but also makes our brains work better. Research shows that when we're interested and engaged in a subject, our brain gets a shot of dopamine—our reward system.[1] That makes us want to keep learning and pushing ourselves.

As noted in Chapter 2, a joyful frame of mind (rather than a fearful one) engages the executive functions of the brain to help tackle tough issues. It also demonstrates increased activity in the creative areas of the brain. In short, we enjoy better thinking for a topic like how to solve climate change now.

What differentiates high performers from others is the amount of extra energy and excitement they bring to their endeavors. People can pursue something for a while, but whether they excel depends upon how it aligns with what gives them joy—what they truly love to do.

Other benefits occur as well. Relationships improve at work and in life. Stress diminishes. Recruitment and retention rates as well as job satisfaction increase.

Doing what we love can be even more fun with other people. This is good news for engaging people in working together for a healthy climate.

You don't need to go to the research labs to confirm these insights. You can turn to your own experience. What do you experience when something is fun?

Lisa Altieri, CEO of BrightAction, has fun using her data science and technical skills. These help her to explore ways to reduce greenhouse gas emissions at a personal level. She and her team discover which sources offer the biggest opportunities. Then she enjoys connecting people with solutions. The result is a robust carbon footprint app, which we'll discuss further in Chapter 6.

Jim Thompson, founder of "THIS! Is What We Did," takes a very different tack in pursuing his interests in advocacy. He describes chanting, singing, and guerrilla theatre as fun activities during demonstrations against banks financing fossil fuels. He reports that these playful actions take attention away from the fear of stepping out and taking a stand. The creative right brain calms the anxious left brain worrying about consequences. He also enjoys the power of poetry to open people's thinking and serve as a tonic for his own brain. Poetry tickles his thinking to see things in new ways.

The point here is not to push you into one form of fun or another, but rather to see the power of fun in energizing your efforts. Fun is your friend in sustaining the energy necessary to achieve meaningful results.

Take a few moments to enter your observations of the impact of fun on your learning, openness to new ideas, energy to follow through, and results.

**Figure 7. Learn from Your Fun**

| Fun experience | Insights you gained | Openness to new ideas | Energy to follow through | Results |
|---|---|---|---|---|
| | | | | |
| | | | | |
| | | | | |

## *Identify what you love*

Since what one person loves to do isn't necessarily what another person enjoys, we each need to examine our own fun factors. Disconnects, frustration, and lack of follow-through usually arise when we've launched off on something that we think we should do or someone has told us to do that doesn't really fit for us.

In coaching, we focus on where people have energy. That's a good sign of whether they are pursuing what they feel they "should" be doing or what they really want to be doing. What energizes you?

Are you wondering how to identify what you truly love to do? Or maybe you'd like some ideas to consider. The table of Activities and Roles below will help stimulate your thinking about what you love to do.

**Examples of Activities and Roles You Might Love**

- *Learning and sharing something new*
  Researcher
  Learner
  Teacher

- *Field trips (in person or online) to see what others are doing*
  Attendee
  Sharing the learning with others
  Organizer

- *Pop-up action events*
  Attendee
  Sharing the news with others
  Organizer

*Identify What You Love*

- *Cooking*
  Planning menus
  Cooking on your own
  Cooking with others

- *Eating*
  Choosing your food
  Experimenting with new choices
  Sharing good discoveries with others

- *Discovering something interesting in nature*
  Taking a hike on your own or with others
  Attending a workshop
  Organizing or leading a workshop

- *Bicycling, walking, skateboarding, or other ways of getting around*
  Solo
  Commute buddies
  Organizer

- *Discussing ideas with others*
  Attendee
  Thought or question prompter
  Organizer

- *Improvisation*
  Audience member
  Player
  Organizer

- *Problem solving*
  Analyzing issues
  Explaining things to other people
  Figuring out new ways to do things

- *Building things*
  Designer
  Planner
  Hands-on construction

- *Shopping*
  Making a list of what's needed
  Looking for best deals
  Telling others about what you got

- *Helping others*
  Providing direct service or support
  Encouraging others to participate
  Organizing a service or event

- *Gardening or landscaping*
  Planning your garden
  Planting trees and/or plants
  Tending and enjoying the garden

- *Encouraging action by legislators, government agencies, businesses, etc.*
  Letter writer
  Speaker at meetings
  Organizer mobilizing people to act

*Identify What You Love*

## *Create your Fun Finder for opportunities*

From analyzing your fun experiences, either on your own or in discussion with a friendly Climate Action Catalyst, you can create a valuable filter for opportunities. Interestingly, the Citizens' Climate Lobby (CCL) uses a Volunteer Inventory Form[2] to find out what people enjoy doing and would like to do. This helps CCL match volunteers' interests with CCL's activities as a nonprofit, nonpartisan, grassroots advocacy climate change organization focused on national policies to address climate change.

Tamara Staton, Greater Pacific Northwest Regional Coordinator, and Education & Resilience Coordinator for CCL, finds that adults are hungry for a sense of lightness and enjoyment in their lives. She finds that play is on par with meditation and mindfulness for effectiveness in releasing creativity and energy.[3] Since some people fear that doing things that are fun for them might somehow discount the gravity of climate change, she enjoys sharing neuroscientific insights about the power of play.

Tamara has found ways to introduce activities like sharing stories. As volunteers have a taste of joyful activities at an in-person Climate Camp, or even virtually through "chatitating" on Zoom, they tap wellsprings of resilience for their advocacy on behalf of CCL. Tamara says, "Having fun helps keep me and our volunteers engaged. Enjoying our time and the actions we take magnifies and amplifies what we accomplish together."

One of the ways to stay refreshed and energized is to explore new types of activities and/or new roles in them. This is your growing edge and a powerful place to play. For example, Tamara is curious about ways to provide Climate Camp-type experiences and wilderness excursions for legislators or government staff members.

What do you enjoy? What additional areas would you like to explore?

**Figure 8. Your Fun Finder**

| What I know I love ||
|---|---|
| Types of activities | Roles to play |
|  |  |
|  |  |
|  |  |
| Additional areas I'd like to explore ||
| Types of activities | Roles to play |
|  |  |
|  |  |

## *Declare to others what you love to do*

When you let other people know the kinds of activities you like and the roles you enjoy, you'll optimize your chances to attract those opportunities.

When I began an intensive focus on climate health, I made a point of talking with trusted colleagues and friends about it. I included my interest in how to have fun taking action on the signature box for my emails. I interviewed people to learn about their thoughts and activities. These efforts quickly created a network of friends and colleagues with useful ideas and insights.

Does focusing on what's fun for you sound too self-indulgent? No. It's not self-indulgent. When you're enjoying what you're doing, you're more likely to keep doing it, and talk about it with other people. And that's what we need for a healthy climate.

Can we count on fun to fuel our success? Yes. It's a path to enlist people in important work. Resolving chronic, long-term issues like climate change requires sustained attention and effort. That requires self-motivation.

The needs for climate health described in the next section offer many ways to play. Collectively, we need to do what needs to be done. The question is, what motivates you?

## *Opportunities for you to enjoy finding what you love*

- Create a Fun Finder for yourself. Write down each of the different types of activities or roles that you enjoy. You can pick roles from personal and work situations. You might want to include activities or roles you'd like to explore, to see if they are enjoyable and meaningful to you. Put each one on a separate index card or piece of paper (or enter it into a digital document). This will help you shuffle and sort them for matchups with climate needs in Chapter 10.

- Enlist your Climate Action Catalyst friends to share what they observe about where they've seen you loving what you do. They can help you expand and confirm your entries.

- Compare what's fun and the kinds of roles that you enjoy with those for other people. Note the differences. When you explore how to connect with the needs for a healthy climate in Chapter 9, you'll want to be sure that each person plays their best position.

When you do what you love, you'll have the self-motivation to do what's needed.

# SECTION II

# CONNECT WITH CLIMATE HEALTH NEEDS

# Connect with Climate Health Needs

There are so many needs to promote climate health that it can be mind-numbing to think about them. Consequently, many people simply freeze and do nothing. So, let's see how you can find climate health needs that interest you.

This section of the book lays out a framework for thinking about the needs. I call it the AAA of Climate Health—Awareness, Actions, and Advocacy. It defines areas in which you can match what you enjoy with what will make a difference.

**Figure 9. AAA of Climate Health Needs**

# Chapter 6

## Promote Awareness
## – *Track Your Carbon Footprints*

*"The first step toward change is awareness."*
~ Nathaniel Branden

Did you know that an estimated 40% of U.S. climate emissions come from basic activities we do every day like using energy at home and how we get around?[1] What's more, our choices about those activities (from whom we buy, how they are created and delivered, etc.) drive 72% of greenhouse gas emissions.[2] So, don't underestimate the importance of what we can do. We have choices. We have power.

People love self-assessments—opportunities to measure where they are and actions they can take. As they say in business, "We manage what we measure." When we measure something, people pay attention to it and often take steps to improve their results.

We need inviting ways to learn where our choices can create the greatest benefits in reducing greenhouse gas emissions. Online measures of carbon footprints provide such a tool.

In assessing our choices, graphical information helps. For example, here's a compelling figure from the Center for Sustainable Solutions at the University of Michigan, about the

GHG emissions from our food choices as part of our personal carbon footprint.[3] How might information like this affect your personal choices?

### Figure 10. Impact of Our Food Choices

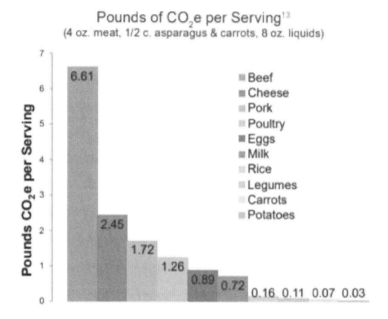

Source: Center for Sustainable Solutions, University of Michigan

If people are to make changes, they need to have and believe the data about their current situation and the potential for improvement.

## *Determine where you are now – your carbon baseline and your choices*

A member of our Rotary club's climate action team asked, "What can we do together for climate health? Our home and work situations differ so much." That's true. You need to start with understanding where you are, and each person, household, or business is different. So, what we can do together is help one another learn about our carbon footprints.

There are several different online carbon footprint applications to consider. The best fit for you depends upon your objectives and your circumstances. Some take a global perspective like the Global Footprint Network (www.footprint calculator.org). Others like www.BrightAction.app seek to give a more fine-tuned assessment of actual greenhouse gas emissions and the impact in emissions and dollars of taking actions.

Our Rotary club's interest in exploring prospects for our members collectively, to become net carbon zero, led us to the BrightAction app. With it, we began to identify options and quantify personal and collective progress to achieve that goal.

For example, our household of three already had some rooftop solar, low-flow water faucets and shower heads to reduce hot water heating, and reduction in our heat thermostat. What more could we do? The app identified reduction in transportation-related emissions as the biggest opportunity, both through offsets in air travel and buying or leasing an all-electric vehicle. Nearer term were additional opportunities to eat lower down the carbon chain, choose full renewable energy from our utility provider or install more solar panels, and switch the last of our light bulbs to LEDs. Each action had a $CO_2$ emissions benefit and a cost or savings associated with it so that we could make choices that fit for us.

Just knowing our $CO_2$ emissions and what we could do to reduce them got members of our household motivated to act. We did some immediate and easy items. We purchased offsets for our air travel (Green-e certified credits from a landfill methane capture project), took advantage of telecommuting, focused on eating lower on the carbon chain with less red meat and more vegetables and fruit each week, and consolidated loads in the washer and dryer. With the completion of these few initial actions, we reduced our $CO_2$ emissions 32% and created a net savings of $728 per year. Delighted with our quick results, we focused on bigger opportunities, like replacing our 15-year-old hybrid vehicle with an all-electric vehicle, and checking out prospects for adding more solar panels to charge it.

**Figure 11. Household Actions Reduce Carbon Footprint**

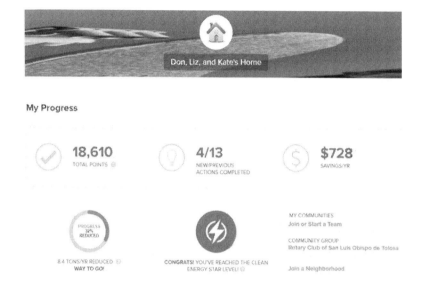

Source: https://BrightAction.app

Other households with different circumstances had individualized carbon footprint calculations and opportunities for actions tailored to their situations.

## Make it a shared project

Most social change strategies benefit greatly from team efforts. For example, Weight Watchers has gained success for its members with weekly weigh-ins. They build accountability and celebrate success. What will help your group gain results for reducing your carbon footprints?

## **Boost other people's awareness of climate issues and opportunities**

You can play a valuable role in helping others become aware of climate issues and their opportunities to make a difference. Marketing experience highlights that people typically need seven to ten exposures to an idea before acting on it. Thus, you need to present opportunities in several different ways and reinforce ideas to ignite change.

Calla Rose Ostrander helped farmers and community members tap science insights and resources with easy-to-understand explanations of regenerative agriculture. Because soils hold tremendous potential to act as a carbon sink and restabilize hydrological cycles, she and her colleagues at Nerds for Earth compiled the economic and environmental benefits of focusing on soil health in agriculture.

Starting in 2016, they gathered research, reports, and other data, and compiled it in an easy-to-access online space. That was early in the soil health movement's efforts to encourage grassroots awareness and education for advocates and locally elected public officials. By identifying needs that front-running organizations had in the soil health space, and building

# Promote Awareness

prototype data models or data visualization, they encouraged ideation, iteration, and conversation.

Nerds for Earth took it another step and reached out to invite people with technical skills to build a State Healthy Soil Policy Map. The crowd-sourced, interactive map focuses attention on important opportunities to enhance soil health through legislative action.

**Figure 12. Using Technical Skills to Boost Public Awareness**

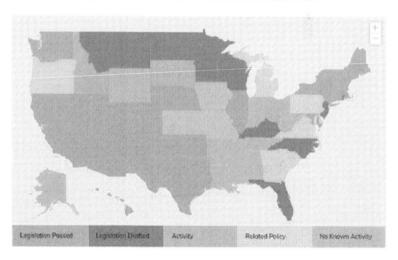

Source: www.nerdsforearth.com

Awareness-building opportunities are especially important in the developing world. As their citizens try to catapult to the envied North American lifestyle, they have especially important choices for the planet. Will their personal paths and choices protect the environment, or compound the problems of waste and GHG emissions?

In response to her concerns for pollution and waste management challenges in rural Argentina, Kelly DeMarco, an occupational therapist and wellness coach from Wisconsin, co-founded The Argentina Bike Project, a 12,000 mile mountain bike trek, to talk about the ugly side of consumerism and the problem of trash. She and her cycling companion, Argentine native Leo Horochowski, used their love for mountain biking, travel, and adventure to embark on an information campaign around the countryside to raise environmental awareness and to spread a message about the value of living with passion and purpose.

With joyfulness in meeting new people, Kelly and Leo made their way into remote villages and dwelling places in all 23 provinces. Their peculiar-looking clothing and gear drew attention wherever they went. This gave them a perfect platform from which to speak to local audiences, TV broadcasts, and newspaper journalists. In addition to training for and cycling 4,300 miles of the journey herself, Kelly assisted with the fundraising, marketing, PR, website management, and sponsorship to pull off this event.

Kelly has carried her enthusiasm forward at home as co-owner and operator of a regenerative farmstead and rustic Italian slow food venue with her family of four in Cascade, Wisconsin.

**Sharing Concern for Climate Health**

Kelly DeMarco and family on the farm

As you think about boosting awareness for yourself and others, use a systems approach. That is, think about how information, awareness, reinforcement, trial, and feedback link together to stimulate meaningful results.

**Figure 13. Stimulating Climate Health Awareness and Action**

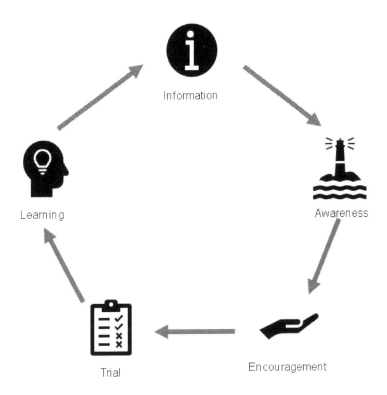

Sustained change often requires multiple rounds on this cycle as people gain manageable bites of information, become aware, receive some reinforcing or corroborating information, put it into practice, and gain feedback.

Where would you like to contribute in this cycle? With whom would you like to connect to fill in the other roles?

## *Opportunities for you to enjoy building awareness*

- Pick a carbon footprint tool and try it out. Look for ones that appeal to your spirit of adventure. If it seems intimidating to get started with an online app, tap someone who can help you. See www.SolveClimateChangeNow.com for additional information about carbon footprint options.
- Identify significant opportunities to reduce your GHG emissions.
- Help other people become aware of their opportunities to boost climate health in ways that they will enjoy.

Awareness stimulates our brains to explore solutions.

# Chapter 7

# Identify Direct Actions – *Live More Richly*

*"Inaction breeds doubt and fear. Action breeds confidence and courage. If you want to conquer fear, do not sit home and think about it. Go out and get busy."*
~ Dale Carnegie

# 7

Start with your personal choices and build out to your home, work, community, and beyond. When you have identified many opportunities, you will feel more encouraged and more empowered that you can make a difference.

When coaching people, I often invite them to take the 100 Resource Challenge. It's a fun exercise to think of all the resources and ways in which you could achieve your objective. Participants find that by the time they've identified 100 resources, they accomplish their objectives.[1] Our brains feed on opportunities and engage the focus and energy needed to realize them.

The discussion below provides some examples of opportunities. Go to www.SolveClimateChangeNow.com to find additional resources.

Based upon your distinctive carbon footprint identified with the kinds of tools discussed in the preceding chapter, you can explore actions that will make a difference, will fit you and your life, and will be fun to do.

## *Opportunities in your personal life*

Opportunities abound in our personal lives. Inspired by the theme of "sufficiency is the new abundance," some people choose to skinny down their personal carbon footprint as discussed in the previous chapter. For example, they look for ways to share rides with other people. They think of carpooling not as drudgery in coordinating with other people, but as a way to expand or deepen their relationships, do a crossword, or simply get some time to rest.

You might consider other opportunities for action, like offsetting the emissions from your travel with donations for the equivalent in planting trees or protecting ancient and endangered forests. There are groups to help you translate your intentions into actions and results.[2]

## *Opportunities at home*

Reduce, reuse, recycle is the mantra here. What are unneeded purchases? For example, do you have appliances or equipment that you can fix and keep working? There are groups like iFixIt (www.iFixIt.com) that enable people to figure out what they can do and how to do it. Instead of wasting time running to the shopping mall or waiting for things to come via resource-draining and $CO_2$-producing shipping, you, your family members, and friends can figure it out and learn something new. Of course, there also are many how-to videos on YouTube and elsewhere on the web.

Our family has fun making our own oat milk. When our daughter thought she was having some nasal congestion from milk products, we explored other beverage choices. As a grain-based beverage rather than a nut-based drink, oat milk consumes less energy and creates a smaller carbon footprint. Indeed, oat milk has a reported carbon footprint that is 28% of cow's milk.[3] (You will also want to consider nutritional and other factors in making your beverage choice.) It's easy and fun to make oat milk. We simply mix one cup of organic oats with four cups of water in the blender and squeeze it through a fine mesh bag. We store the result in a reusable glass jar in the refrigerator. Now, we have oat milk whenever we need it. In fact, it's become a fun father-daughter project.

## Opportunities at work

Does your workplace provide incentives for saving energy and reducing the production of greenhouse gases? If not, you could gather a group and start a program. Recycling is one step. If you track your carbon footprints, you can identify and celebrate the reductions and cost savings you gain. Challenge your employer to match the percentage of GHG emissions you are saving at home.

The important thing is to use your power as employees, to call for action by your employers. Many employers welcome these efforts because they've learned that employee engagement, as well as environmental and bottom-line benefits, can flow from aligning with their employees.

Organizations like the Center for Climate and Energy Solutions (www.c2es.org, successor to the Pew Center on Global Climate Change) identify options and resources.

## *Opportunities in your community*

Think about active social services organizations and community clubs with whom you can partner. Many groups like to be of service, engage, and support a good future for generations to come. Seek them out.

For example, we organized a Climate Action Team in our local Rotary club. Since Rotary International has made environmental issues a major focus, there is lots of information sharing among clubs, and support for action. The civic leaders in our club chose to focus on reducing our carbon footprints, planting trees, and enhancing soil health and capacity to convert $CO_2$ into living carbon. Our stretch goal is to become net carbon zero as a club. We want to be part of the solution with our local government, which has a similar goal of becoming net carbon zero.

## Opportunities beyond

One of the exciting things about focusing on greenhouse gas reductions and mitigation efforts is that actions anywhere benefit all of us. So, enlightened self-interest invites us to broaden our horizons.

For example, contributing to the restoration of mangroves in coastal African communities not only enhances carbon sequestration; it also provides protection from the ravages of storms and flooding from climate change that plague low-lying, often low-income areas. Sometimes a dollar invested in such economies goes farther toward denting climate change for us all than a dollar in our higher-cost areas.

## Align with nature to solve our climate crisis – Attraction in action

For eons, nature has been converting $CO_2$ in the atmosphere into living carbon in trees and plants. All was fine until we overloaded the system with man-made $CO_2$ and reduced the trees and plants available to convert it.

We can turn this around. Steps to reduce our carbon footprints, protect our forests, and restore soil health offer ways to get us there.

Working with how nature creates climate health provides interesting and enjoyable opportunities to learn. We can discover more about the way ecosystems work. We gain insights for how to live more satisfyingly in harmony with the environment. And, we have something lasting to share with our children and grandchildren.

Initiatives that align with nature have a natural attraction to people. They underscore the abundance of resources we have available around us for this effort.

Here's an example of the magnetic and organic ways in which natural solutions can unfold. Your needs and interests may differ. The important take-away is how the attraction model of sharing hopes, inviting people and resources to join you, and celebrating success generates results. As you read this example, I hope it inspires you to expand your horizons of what's possible.

**Attracting People and Resources to Contribute to Climate Health**

Some people in our church (St. Benedict's Episcopal Church) wanted to beautify the property with more trees. We had clergy who shared stories that churches during the Middle Ages planted trees on their properties to restore the forests decimated by the ravages of conflict and poverty. Others saw tree planting as an opportunity to sequester more carbon. As noted in Chapter 2, welcoming multiple, complementary objectives enabled more people to get excited about the effort. A generous community member donated funds for landscape planning to identify possible locations and types of trees.

Not surprisingly, the plans for tree planting also raised fears. We're a small congregation and didn't have the money in our budget for the trees and their planting, much less for the irrigation and tending needed in the crucial first three years. Many older members couldn't put their backs to the shovels. The irrigation system was old and broken down. Estimates came in at around $7,000 from contractors to fix it. People became discouraged.

Instead of giving up, a group persevered. Members put the word out that our church wanted to be a resource for the community, help beautify it, and contribute to climate health. This attracted the attention of the local garden club. Its leaders knew of a special tree planting grant coming to a local environmental group. In response to wildfires, the state was distributing funds to purchase and pay for planting trees. Our church's land, which could accept 30 large, climate-hardy and carbon-sequestering trees, offered an attractive opportunity.

From previous decades of farming that occurred on the property, the soil had become little more than hardened clay with weeds and ground squirrels—another challenge. What could we do about that? Through the local community college, several of us attended Saturday workshops on regenerative agriculture. We learned that nurturing naturally occurring organisms could make the soil more fertile and multiply the ability of the trees and other plants to convert $CO_2$ into living carbon organisms that would stay in the ground. We even saw how to make a bioreactor to concentrate the natural biology. It requires about 1500 pounds of dried leaves, horse manure, and wood chips in a breathable bin where everything is kept moist, with access to air for 12–14 months while the organisms in the air and in the plant material multiply.

The bioreactor required about $300 in materials, plus labor to gather the ingredients and assemble it. Fortunately, Tim LaSalle, an expert on regenerative agriculture and the nurturing soil health, lived only about 20 miles away. He offered to share some materials and guide us in building the bioreactor. (You can learn how to build a simple bioreactor on the web.[4] ) We went to the church leaders. Thankfully, they approved the project, even though we didn't have the money and labor. It was a leap of faith. Delightfully, a member, who hadn't participated in any of the planning, was taken by the idea and the enthusiasm it generated. She donated the money needed for materials.

*Identify Direct Actions*

What about the labor? We approached a local Rotary club. In 2020, Rotary International identified the environment as one of its primary areas of focus. Our local club had recently joined the Rotary Climate Action Network of over 400 Rotary clubs around the world dedicated to working for climate health. As noted earlier, the local Rotary club shared an interest in tree planting and building bioreactors to enhance soils. A group from the club agreed to work with able-bodied members of the congregation and the local garden club to build the bioreactor. This became the club's opportunity to learn by doing. Together, we built the bioreactor on a Saturday morning. For immediate results for our tree planting, we borrowed a quantity of material from the nearby City Farm, to be repaid when the product of the church's bioreactor became available in a year.

But what about the $7,000 for the irrigation fixes and extension? With the COVID pandemic still underway and many people doing home improvements, irrigation contractors were not only costly but also difficult to get. Were we stuck? The local garden club's connections with another Rotary club garnered $500 for the tree project. Then, another church member committed $5,000 for help with the irrigation requirements. Still, we were short of the total needed, and without any contractor to do the work even if we got the money.

Just as despair was about to overtake us, a solution arose. A member of the congregation had a daughter returning to the area. She had experience in landscape work. He offered to volunteer his labor and do the job with his daughter's paid labor, for no more than the amount raised.

Two months later, ECOSLO (our local environmental group), the Los Osos Garden Club, and members of our church planted the trees with an irrigation system in place to tend them.

**Community Comes Together for Climate Health**

Community members from St. Benedict's, ECOSLO, and Los Osos Garden Club after successful tree planting

*Identify Direct Actions*

Some people might discount this story. *"Could all of those circumstances line up in other situations?" "Are you suggesting that we should rely upon efforts like this to help get us to climate health?"*

Yes, they can. Yes, YOU can!

For over 25 years, I've worked with groups to solve the unsolvable.[5] They've built schools when bond issues had failed twice before. Communities have gained agreement on shared goals and have accomplished them even when they previously were at loggerheads.

The process is the same. Focus on your hopes. (Recall Chapter 2.) Listen and learn from others. Improvise so that obstacles become steppingstones to success. Let people know what you hope to accomplish in order to attract their aspirations and resources to join you.

Each group, business, or community will need to discern where it's called to act for climate health. As you go forward together, you will be creating deep bonds among your members and with others who will sustain you in the ongoing work and joys.

Where would you like to sink your roots and make a difference?

The following are a few ideas about action opportunities.

## Examples of Personal Actions

| *Types of actions* | *Examples* |
|---|---|
| Personal life | Rideshare, reduce car trips, lease or buy electric. |
| Home life | Eat foods that generate less GHG emissions; reduce, recycle, reuse. |
| Work | Create/join a team to reduce your business' carbon footprint. |
| Community | Join with a group to plant trees; improve soil health. |
| Beyond | Donate to preserve ancient and endangered forests, restore mangroves, and other efforts for climate health. |

Of course, advocacy efforts for climate health are very important. We'll address advocacy in the next chapter. It's important, however, to ground our advocacy in the integrity of having taken personal action ourselves.

You can find more ideas and references to action resources at www.SolveClimateChangeNow.com. You are also welcome to share your suggestions for other actions you've found to be effective and fun.

*Identify Direct Actions*

"Grace happens when we act with others
on behalf of our world."
~ Joanna Macy

**Opportunities for you to enjoy direct actions**

- Pick one or more domains (personal, home, work, community, and beyond) that appeal to you.
- Identify actions that sound like they'd be valuable and fun to pursue.
- Choose some actions to get started.
- Look for ways to align with nature to promote climate health together.

Imagine what can be accomplished when people say "yes" to action.

## Chapter 8

# Find Advocacy Opportunities
# – *Share Your Story*

*"Never doubt that a small group of thoughtful, committed citizens can change the world; indeed, it's the only thing that ever has."*
~ Margaret Mead

Some people start with advocacy. Most people come to it after boosting their own awareness and taking actions. They understand the challenges, experience the joy of making a personal difference, and then want to magnify that difference in public policies.

Are you wanting to have a bigger impact, but you aren't sure about how to go about it or which efforts to join? Or maybe you know what needs to be done, but you feel reluctant to put yourself out there? I appreciate those feelings. For many years, I felt more comfortable being like "Switzerland," not taking sides in disputes. In part, that fit with my work as an independent facilitator seeking to bring diverse interests together to solve tough issues. I didn't want to be seen as favoring one direction over another. This also fed my desire for people to like me. I wasn't rocking anyone's boat.

The shift for me came from several sources. First, the climate challenges we face simply can't be ignored. I can't be inactive and still look in the eyes of our daughter or our friends' children and grandchildren. There's too much at stake. Wittingly or unwittingly, it's largely the Boomer generation that has created the problem. We need to take responsibility for helping to solve it. Also, while my advocacy may alienate some people, I've found much deeper and much richer relationships with people who are ready to act. I see people at shops and social gatherings who thank us for taking action and pushing for change, and who want to join. It's rewarding helping people see that they can make a difference and have fun doing it.

What motivates your inklings toward advocacy? Who and what do you care about enough to take a stand for climate health?

## *Target a tipping point – It's lower than you think*

We don't need to get everyone to agree in order to turn advocacy into results. In fact, as noted earlier, trying to get large numbers of people to agree on the same way of thinking stymies rather than encourages action. While there may be some differing views about climate in my Rotary club, we can all enjoy planting trees together. From planting trees together, we can explore ways to magnify that effort on regional, national, and international levels.

Research shows that major social movements typically require only about 25% of the population to reach a tipping point.[1] Why? Because a few dedicated people far outweigh the ranks of those who are indifferent. If people are willing to reinforce their advocacy with non-violent civil disobedience, studies show that the engagement level may be as low as 3.5% of the population.²

The critical factor for advocacy to succeed is to keep the advocacy invitational. If we bear down on people or, worse, cajole them to engage in a specific form of advocacy, we create resistance.

## *Explore a spectrum of advocacy opportunities*

Some people shy away from advocacy because it feels too confrontational for them. Their amygdalae go on alert and they want to flee. In fact, there are many different forms of advocacy. You might think of it as a spectrum. The Rev. Dr. Margaret Bullitt-Jonas, a long-time advocate for climate health, distinguishes between being political (advocating for policies and actions) and being partisan (advocating for specific parties or candidates). With this distinction in mind, even non-profit organizations (community service organizations, faith-based groups, good-government organizations, etc.) can be political.

Here are Bullitt-Jonas's six forms of advocacy. We can:
- Vote
- Contact decision makers
- Testify at public hearings
- Get published (letters to the editor, opinion pieces, etc.)
- Pray
- Engage in direct action.

You can hear her description of each of these, with examples, in the video, "How Are We Called to Be Public Advocates for Climate Health?"[3]

You can begin exercising your advocacy skills at any level. As you advocate in one form or forum, you'll strengthen your advocacy muscle and gain confidence to expand your efforts. You'll also attract others to join you.

**Share your stories**

Advocacy experts underscore the value of sharing stories about why climate health is important to you. Even if legislators may have a different point of view, your story has integrity and value.

For example, if you are opposing drilling or fracking for oil and gas, you can speak in concrete terms about your concerns for how those actions affect tangible things of importance to you. Maybe it's drinking water. Or perhaps it's air quality, soil contamination, or impact on wildlife and recreation that you enjoy. No one can fault you for sharing the truth of your experience and what you value.

### Be joyful

Here's an interesting story that Margaret Bullitt-Jonas shared with me.

"A few weeks ago, I took part in a sit-in at the office building of Enbridge Corporation in Waltham, Massachusetts, the outfit behind Line 3 and many other dirty energy projects. The event included lots of singing—to me, that is FUN—and we were accompanied by a tuba and two drummers, so we made a lot of noise. That was fun.

"Fun includes making music, making banners, making larger-than-life puppets, performing a skit—all of them ways to show that the human spirit refuses to be silenced. Art, dance, and music keep our spirits alive. These are important elements that make a climate action or other protest feel fun, life-giving, and energizing."[4]

How would you like to make a joyful noise or otherwise stimulate interest in climate health?

### Consider ways to join with others

If all of this sounds too challenging to do, the good news is that you don't need to do it alone. You can find a group to join and benefit from their expertise in effective advocacy. Which perspectives and organizations appeal to you? When you feel in alignment with a group and its approach, it's a lot easier to work together.

Jim Thompson founded "THIS! Is What We Did," to mobilize everyday people to grow a movement strong enough to break the power of the fossil fuel industry, and to stimulate the effective, drastic action needed to spur climate justice and give future generations a chance for a decent life.[5] The group offers a welcoming community, promotes climate change literacy, and pursues easy-access on-ramps to effective action. These include mobilizing demonstrations at big banks that finance fossil fuels, and helping people move their money elsewhere.

"THIS! Is What We Did" Call to Action

Source: www.ThisIsWhatWeDid.org

A local example for me is participation in the San Luis Obispo Climate Coalition. It's an action-oriented group that strives to expand carbon-free energy. The Coalition's engagement with local elected officials and service groups, such

*Find Advocacy Opportunities*

as our Rotary club, target ways we can work together toward shared goals like becoming a net carbon zero community.

My philosophy of living simply so that others can simply live, draws me to broader advocacy interests. This arises from a spiritual philosophy of sacrificial giving; that is, trying to share with others even if it means less for me. Truthfully, though, when I give to others, I become less focused on what I have or don't have. From this foundation, I've been drawn to groups like Interfaith Power & Light, a cross denominational non-profit that seeks to mobilize people of faith. I also connect with the Episcopal Public Policy Network, which tracks legislation and offers easy-to-implement action alerts to express my views to elected officials. Both offer excellent guidance on how to be an effective advocate within each person's comfort zone or the growing edge of their comfort zones.

Go to www.SolveClimateChangeNow.com to find a list of advocacy organizations. Find groups that resonate with you. Please also submit the names and contact information for other advocacy groups that you suggest sharing with interested readers.

## *Advocacy opportunities for you to enjoy*

- Jot down the interests and beliefs that motivate your interest in standing for climate health.
- Pick an advocacy organization or type of organization you'd like to explore.
- Explore some steps to strengthen your advocacy muscle and make a difference.

When we add our voice with others, we multiply our impact.

# Chapter 9

# Select Needs That Interest You – *Apply Your Needs Filter*

*"Life presents many choices;
the choices we make determine our future."*
~ Catherine Pulsifer

With your understanding of the various needs to bring forth a healthy climate, which needs appeal to you? Where do you feel drawn to serve?

Since there are so many needs and opportunities to serve, it can feel overwhelming. So, you'll want some way to sort through them. Here is a process that I found helpful.

## *Create and apply your Needs Filter*

When I faced choosing needs, I developed some criteria to apply. I offer my story here to stimulate your thinking about how to create your own criteria.

My Needs Filter included a compelling social cause, people with whom I'd enjoy learning and working, and where my distinctive strengths are needed. I literally drew the following diagram for each opportunity I considered.

## Figure 14. Example of a Needs Filter

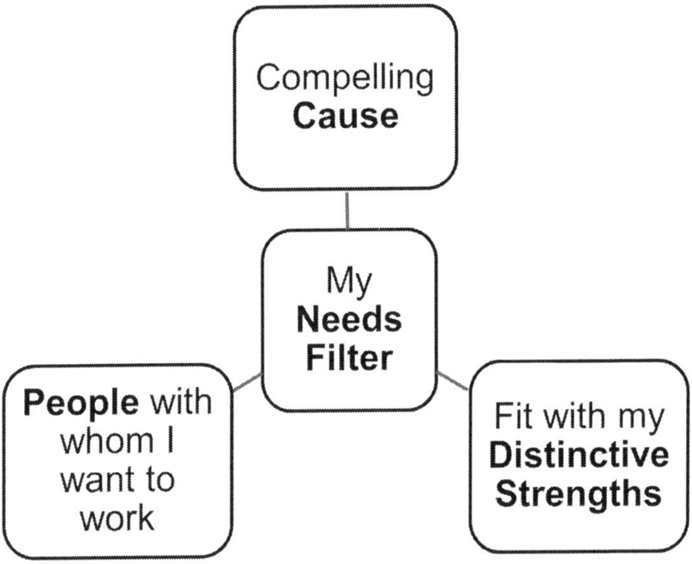

```
        ┌──────────────┐
        │  Compelling  │
        │    Cause     │
        └──────────────┘
                │
        ┌──────────────┐
        │     My       │
        │   Needs      │
        │   Filter     │
        └──────────────┘
         /            \
┌──────────────┐    ┌──────────────┐
│ People with  │    │ Fit with my  │
│   whom I     │    │ Distinctive  │
│   want to    │    │  Strengths   │
│    work      │    │              │
└──────────────┘    └──────────────┘
```

I knew that I needed a clear cause, something compelling to accomplish. This would sustain my interest and commitment. Part of having fun is having people with whom I enjoy learning and working. I'd seen what a difference that had made in the success of my entrepreneurial endeavors and volunteer service. Finally, I checked whether the situation needed my distinctive strengths. With so many things needing to be done, I didn't want to duplicate what others are doing or even compete with people more qualified and available to do the work. Since my distinctive strengths often closely align with the kinds of roles I enjoy doing, that is a good factor to tie back to my Fun Filter (Chapter 5).

The opportunity areas for me included finding people who wanted to act to measure and reduce their carbon footprints. At about the same time, a local Rotary club encouraged me to join them. Serendipitously, Rotary International had just declared the environment as a new major cause, adding to its social service initiatives like eliminating polio. I wondered if maybe climate health could be, like a previous campaign to eradicate polio, a rallying point for Rotary.

I joined the Rotary club and found a core group of can-do people who were eager for the opportunity and demonstrated that they knew how to have fun being of service. Members had a genuine joy in doing their service projects. I could learn from their effective ways of engaging and organizing members and people across the community. My budding knowledge about climate change, and my ability to develop a conceptual framework and set of strategies to implement them, tapped my distinctive strengths and complemented their strengths. With that, we had a promising opportunity that fit my Needs Filter criteria.

Similarly, I found appealing opportunities in our church. I joined the previously established Earthcare Team of members with an interest in how communities can support climate health. They had completed a book study guide on Jim Antal's book, *Climate Church, Climate World: How People of Faith Must Work for Change*.[1] Together, we turned to "how to mobilize people of faith to take action." We decided to develop the "Beginner's Guide to Creation Care and the Climate Crisis."[2] We needed someone to pull the five-session course together and moderate the video presentations for each. Since I had developed and facilitated large-scale coaching program webinars for the International City/County Management Association, I had the technical expertise and presentation skills for the role.

Since I was pursuing what I love to do, and following criteria that aligned with what's important to me, these opportunities unfolded in a natural way. I want you to enjoy a similar experience. I hope my personal examples encourage you to find the path that is calling to you. You don't have to figure everything out in advance to get started. Explore potential needs. See which ones attract and sustain your interest.

## *Communicate where you want to play*

To find the opportunities that you want, you similarly need to let people know what you enjoy doing and what the needs are that interest you. You will quickly attract multiple opportunities and have a reason to apply your Fun Filter and Needs Filter to find what fits for you.

Be clear about what you want. Tell more people about your interest. You will discover fulfilling opportunities.

For example, I shared my interest with my network of Master Certified Coaches. From these efforts, the coaching coordinator for the Climate Breakthrough Project contacted me about a pro bono coaching opportunity with the leader of Canopy, one of its grant recipients. Canopy works to divert the fashion and packaging industries from using ancient and endangered forests for the fibers in their products. They have a promising set of next generation solutions. (See www.CanopyPlanet.org.) This sounded compelling to me. Nicole Rycroft, Canopy's founder and executive director, and I quickly developed a strong rapport. The organization valued my distinctive experience as a social and business entrepreneur and venture investor.

Each of these situations has been a delight and blessing to me. While each of us is different, I'm confident that you can find opportunities that are similarly rewarding for you. Just have a filter that works for you, communicate your interests clearly, and pursue and explore what appeals to you.

If you feel stuck in finding connections, tap a family member, friend, or colleague to brainstorm with you. If you want some additional support, check the Climate Action Catalyst resources at www.SolveClimateChangeNow.com.

Remember, it's not how big you start. It's how big you play. Look for fun ways for you to enter the game. Each effort, no matter its breadth or depth, adds to a collective solution.

## *Opportunities for you to enjoy picking needs that interest you*

- Write down areas of need within Awareness, Actions, and Advocacy that appear most compelling to you. These can be within personal, household, work, and community situations. Put each one on a separate index card, piece of paper, or entry in a digital document. This way, you can shuffle and sort needs that interest you. (See Chapter 10.)

- Decide upon the criteria that are relevant for your Needs Filter, and apply them to the ideas that interest you. Enlist your Climate Action Catalyst friends to share what they might suggest about your list.

- Compare your list with those from other people. Look for some common areas of interest. Also note the differences. Discover how you can complement one another's interests.

Visualize yourself making a difference for climate health.

# SECTION III

# ENJOY YOUR CLIMATE SWEET SPOT FOR A HEALTHY PLANET

# Enjoy Your Climate Sweet Spot for a Health Planet

Now's the time to pull your plans together and put them into action. This is the especially fun and fulfilling part. You can do this on your own or have even more fun doing it with other people in your family, work, or community.

**Figure 15. Matching What You Love and Climate Needs**

# Chapter 10

## Create Your Action Portfolio
## – *Celebrate and Share Your Success*

*"Success is not the key to happiness.
Happiness is the key to success. If you love what
you are doing, you will be successful."*
~ Albert Schweitzer

# 10

Thank you for the attention and care you've brought to exploring what you love to do and the needs to promote climate health. You've done the spade work. What are you ready to plant, nurture, and grow? Enjoy putting your picture together and finding satisfaction from making a difference for yourself and the planet.

## *Put your cards on the table*

Take the cards from Chapter 5 on which you wrote what are fun activities and roles for you, and put them on the left side of a board or table. Then, take the cards with the needs that appealed to you from Chapter 9, and put those on the right side. Enjoy looking at them and imagining what you might do.

If you are doing this with others, have each person explain their cards. Where do you have similar interests? Where do you differ? You can still play together even if you have different interests in activities and roles or in needs. In fact, some of the most powerful teams are those that collaborate with complementary interests.

### *Explore matchups of what's fun for you and serves climate health needs*

Have you ever wondered how tech entrepreneurs come up with new apps and other exciting products? They don't start from scratch imagining something. More often, they use matchups of bits and pieces of things that they've already made. This is also how people come up with new ideas. They find ways of mixing and matching ideas and experiences that they already have, and build on them.

You can take a similar approach in developing your opportunities. Pick up a card from the activities and roles category and one or more cards from the needs category. Brainstorm different ways in which you could put those together and make a fulfilling difference for climate health. You'll find that doing this with someone else will multiply the opportunities for you. Other people will not have implicit or explicit constraints that you've put on yourself. Like, "Oh, I can't do that." Rather, they might offer, "How about if you tried this?" And as you brainstorm, remember the improv rule: avoid "no" and "yes, but...." Instead, use "yes, and...." As described in Chapter 2.

You can start in your own physical or virtual neighborhood. Here's what Calla Rose Ostrander, from Nerds for Earth, suggests from her work as a catalyst connecting talent with communities in Colorado, Texas, New Mexico, and California.

Do what you can with what you have where you are. For example, if you are really good at coding and want to do something for climate health, you don't need to be a climate expert. Seek out people in your neighborhood or the larger virtual community and see what their needs are. Maybe they would like an app for tree health, or some other measure of climate health that might spur awareness, action, and advocacy. Be interested and curious about them. Get to know them.

Fighting climate change at large isn't necessarily fun. But getting connected with neighbors and solving issues is fun. Play for a purpose. Enjoy playing with people who want to play with you and with whom you want to play.[1]

How might you match up what you love doing with what's needed to solve climate change?

## Create your climate action portfolio – Where and how you want to play

Perhaps, you've thought of multiple things that you'd like to do. If so, you may be wondering how to balance the opportunities you want to pursue. How can you make the biggest impact, with the least amount of effort, and do it in a way that doesn't burn you out?

One visual way to do this is to use the Opportunity Sorter.[2] First, rate how beneficial each opportunity is (high or low). Then determine what level of resources—time, effort, and money—would be required to pursue it. Lay out the opportunities on the "Opportunity Sorter" grid. (See diagram.)

### Figure 16. Opportunity Sorter

|  | Resources – low | Resources -- high |
|---|---|---|
| Benefit high | ★<br><br>Go for these. | ?<br><br>Choose selectively. |
| Benefit low | √<br><br>Give low priority. | X<br><br>Avoid these. |

Those opportunities that offer high benefits with low resources are represented by a star. They provide the "biggest bang for the buck." Opportunities with high benefit but high resources are shown as question marks. Is the return great enough to warrant the effort? Most individuals, teams, or organizations can support only one or two question marks at a time. There's just too much effort involved. Sometimes, participants can identify ways to pursue the question marks with less effort so that they become stars.

The check marks are like popcorn—low benefit but low cost. These are OK to pursue if they don't distract from the stars. Finally, the X's are losers whose high resource requirements outweigh the benefits.

Select the best combination of opportunities to accomplish what you hope to achieve.

You may want to do some of the things that show up as check marks while you are lining up efforts to accomplish the stars.

## *Enjoy rallying a group*

You can do all of the things in this book on your own, but you're likely to enjoy them more by doing them with other people. Here are a few guidelines for engaging a group of people.

*Self-organize to let people do what they want.* When you follow the process outlined, to have each person identify their preferred activities and roles and the needs that appeal to them, you engage their self-motivation. People don't feel coerced into someone else's preferences. Plus, you'll learn useful information about one another and how to capitalize on your strengths and interests.

*Offer appetizers to encourage people to sample and try different opportunities.* Few people will order a full course dinner with dishes that they've never tried before. They need to sample something to see what they like. You can follow a similar path in offering an opportunity to showcase possibilities and easy ways to get involved in initiatives that appear desirable.

*Gather feedback to boost results.* See what can be done to make the opportunities more attractive. You can brainstorm ways to reduce the effort or resources needed to get a bigger return. You'll discover how to make your efforts irresistibly attractive and, thereby, sustainable.

*Be Climate Action Catalysts for one another.* Pair up or set up small groups. Invite each person to be a catalyst for one another. As in chemistry, a catalyst stimulates a reaction or accelerates results. This occurs without taking over another person's initiative and responsibility.

*Engage a Climate Coach.* There's a growing cadre of professional career and business coaches that want to help encourage and support people to take action for climate health. For example, the Climate Coaching Alliance has been active in mobilizing coaches. You can find such resources available on a pro bono or fee basis at www.SolveClimateChangeNow.com.

*Create Your Action Portfolio*

## *Map your path to success*

What's your plan to accomplish the actions you've chosen? You will get there more quickly with a map. Some people call it a results chain. It's a depiction of how you see getting from where you are to where you want to be through your activities. It truly helps to make it visual. As a sample, here's the picture we developed to show our Rotary club how our different initiatives can weave together to realize the big dream of becoming net carbon zero.

**Figure 17. Example of a Map to Results
– Rotary de Tolosa Climate Action Team**

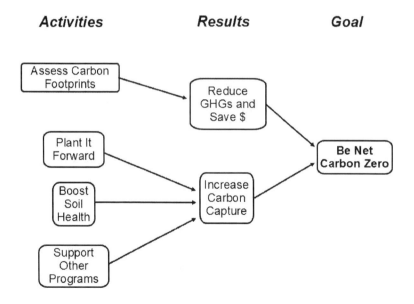

We assess our carbon footprints through the carbon footprint app described in Chapter 6. That leads to our members' actions to reduce their greenhouse gas emissions and save money. In a parallel track, we plant trees through our Plant It Forward program and build bioreactors to enrich the soils with naturally occurring fungi. These convert $CO_2$ into living carbon in the trees and plants. We complement our direct efforts with collaborations with other programs (nationally and internationally), including those from other Rotary clubs. As we scale these efforts, we've calculated that we can achieve our goal for members to become net carbon zero.

As we shared this roadmap with people in our community, we attracted more and more people and organizations that wanted to participate. The city had targeted becoming net carbon zero in its operations by 2030, and finding ways for the whole community to be net carbon neutral by 2035. Its elected officials and staff were eager to collaborate with a community group sharing the objective. They wanted to showcase our efforts to attract additional community organizations to step forward.

Even if you don't have a supportive local government or set of collaborating organizations, you can be the catalyst for action. Your initiative can get people talking, acting, and advocating for climate health.

## Get started with a few early wins

What are some actions you can take to get traction? Identify some early wins for yourself and your team if you are working as a group. As you take action and gain results, you not only make a difference but also strengthen the position of integrity from which you can more strongly and effectively attract the interest and actions of others.

Target your early wins for the next 30, 60, and 90 days.

**Figure 18. Create Your Action Plan**

| Personal actions | When | How they'll make a difference |
|---|---|---|
|  |  |  |
|  |  |  |
| Team or group actions |  |  |
|  |  |  |
|  |  |  |
|  |  |  |

## Celebrate and share your successes

The climate challenges we face require sustained action from as many people, organizations, and governments as possible. Celebration and sharing provide the fuel for self-motivation. Here are some steps to follow.

- Track your results. Measuring your results and the difference you make will keep your mind engaged.

- Amplify what's working. Note where you are attracting more interest, resources, and results. Double down on those opportunities.
- Share what you are doing with others.

## *Opportunities to enjoy your climate sweet spot*

- Engage your Climate Action Catalyst friends to brainstorm the portfolio of opportunities you'd like to pursue.

- Use the Opportunity Sorter to sort through the opportunities that arise. Look for ways that you can boost the benefits and/or reduce the resources to pursue them. What looks like a focused portfolio for you, your family, your business, or your community organization to pursue?

- Track your progress. As you learn about what works and what doesn't, tune up your portfolio. With choices, you won't feel stuck working on something that looked promising but didn't pan out.

- Celebrate and share your successes.

Enjoy taking action.

You can solve climate change now.

## Try the Quick Start Guide for Climate Action Conversations

In order to mobilize the people needed to solve climate change now, we need to reach many people who haven't taken action already. How can you do that in ways that feel inviting to you and to the person you are contacting? Here are some easy steps.

***Begin with a warm welcome***

Use the attraction model and ask for permission to discuss the topic. Most people don't want an uninvited proselytizer. So, you'll want to ask them if they are available to discuss the topic, and if now's a good time.

Come from a place of service. That is, you are trying to support them rather than push your agenda.

Share something of your story. People like to hear short, engaging stories. Think about what you can share from your experience.

Start with a person who you think will be open to the discussion; for example, a family member, friend, or interested coworker. Build your skills to engage others.

## *Consider this sample introductory dialogue*

Just to make the example a little more challenging, let's imagine that you'd like to invite a supervisor or a client you serve. Here's how you might approach your Climate Action Catalyst role (CAC). Let's call the person you are meeting Brenda.

CAC:   Hi, Brenda. I'd like to share with you something I'm enjoying about helping solve climate change, and think would be of interest to you as well. Would you have a few minutes?

Brenda: Well, maybe not now, because I'm just racing to another meeting. What do you have in mind?

CAC:   I've been reading an interesting book about how to find joy in addressing climate change. Through it, I've learned how to help myself and others match up what we enjoy doing with climate needs. I've been having fun making a difference. Would you like to have a short Climate Action Conversation with me later, and see if it appeals to you?

Brenda: Sure. I'll give it a try.

CAC: Super. How about getting together after our team meeting next Thursday?

Brenda: That'll be fine.

OK. You've got a willing prospect for a Climate Action Conversation.

## *Enjoy a sample Climate Action Conversation*

The following quick start Climate Action Conversation illustrates ways to incorporate the key elements of *Solve Climate Change Now* in dialogue. This is a transcript from an actual unrehearsed conversation between Seth Bush and me, where I served as a Climate Action Catalyst. Seth is a professional coach who was interested in experiencing this conversation to see how he might use this format to attract others to take climate actions. The conversation was six and a half minutes, something you can fit easily into your schedule.

You'll note that in my Climate Action Catalyst role, I'm asking a set of questions, listening to Seth's responses, and reflecting back the gist of what he says. This approach not only assures Seth that he's been heard, but it also enables him to hear how his thinking sounds. When people declare their ideas and intentions to someone else, they also become more likely to follow through.

## Seth's Climate Action Conversation

Seth Bush and Don Maruska in Climate Action Conversation

Don: Seth, as you think about how you can be more engaged in climate issues yourself, or get your organization more involved, what are some of your hopes for a healthy climate? What's animating your interest?

Seth: I live in a neighborhood where climate doesn't feel super real to some people. But it's actually under the surface all of the time. We've been dealing with more intense storms and a lot of flooding and power outages as a result. I've seen some of my neighbors really suffering from that. I'd love some more ways to be helping in tangible, simple ways. I hear a lot of talk about what's happening far away regarding climate, but I'm just hoping for a healthy neighborhood. I'm active in local government and want to see climate addressed here.

Don: This sounds very personal and a neighborhood kind of issue. You're looking at how to make a difference and what ways there are to do that.

Seth: Yes.

Don: OK, great. Next, we are going to look for where the kinds of activities and roles you enjoy doing match the climate health needs in your neighborhood area of interest. This will be your climate sweet spot. It's where you will have sustainable energy to pursue your efforts because your efforts come from a place of joy or fun for you. OK?

Seth: I like joy.

Don: Good. Let's move to this question. What kinds of roles or activities do you enjoy?

Seth: Gardening, biking, and learning about new music with friends are probably the three activities that come to mind first and are really fun. And sipping scotch and listening to a new album with a dear friend.

Don: OK. So, gardening, biking, being with friends, enjoying music, and enjoying some scotch and music together. Note: we have a list of activities and roles to stimulate ideas. (See Chapter 5.) However, since your responses came so quickly to your mind, we can move immediately to the next question. What kind of climate needs appeal to you? They could be building people's awareness of climate change, their carbon footprints, and available solutions. It could be taking direct actions at home, work, or in the community. Or it could be advocacy where you want to influence policies and decisions. You can play in more than one of these areas, but for the moment, which one of these jumps out to you?

Seth: Well, I'm an action person. So, action just really appeals to me first.

Don: OK. Action is it for you. What we are going to do now is look at how you can match up what's fun for you, with climate needs. When you think about biking, gardening, being with friends, music, and your interest in action to enhance your neighborhood, what are some matches that pop into your mind?

Seth: Just so you know, I only have about two minutes before I need to go to my next meeting.

Don: That's fine. We're close to the end.

Seth: What jumps to mind first is that I do love gardening. It seems like a great way to engage through action with some other people in my neighborhood, to improve some of our green spaces so they don't get paved or developed. This would enable us to manage some of our storm water problems. Yes, it might be kind of fun to garden with neighbors.

Don: Alright. Super. You've found your climate sweet spot. I'm excited about your plans. Before we close, do you feel that this is the kind of conversation you could have with other people in your organization or with the clients you serve?

Seth: Yes. I am appreciating the overlap for people, between the needs they're seeing and what lights them up. Both in this conversation and elsewhere in my life, that's the heart of it. I appreciate the simplicity of how this conversation unfolds. It's not complicated. It's really simple questions: What do you love? What are the needs? Where are the matchups?

Don: Thanks, Seth. I've enjoyed learning about you and your interests. It's great that you've identified immediate opportunities to take action and have fun doing it.

When I'm guiding workshops on *Solve Climate Change Now*, I model this conversation with a volunteer like Seth. The remainder of the participants observe to get a feel for how it works. While they are observing, I encourage them to be thinking of creative ideas to match up the interests and needs they hear. The observers enjoy the fun of developing possibilities for the person in the conversation. This not only expands the array of options for that person to consider, but it also motivates the observers about the potential of this type of conversation for themselves.

As an observer commented: "I love when you said, 'Do what you love and match it with your actions.' It's simple, but it's so true." Another said, "I want to talk about my sweet spot, and I think everyone does if you can help them tap into it."

Here's a short summary of the key questions for a Climate Action Conversation:

- What are your hopes for a healthy climate? Why are they important to you?
- What roles or activities do you enjoy?
- Which climate needs appeal to you (awareness, actions, advocacy)?
- What are the climate sweet spots, where what you love and climate needs meet? (Would you like some additional ideas that come to my mind?)
- What could you do now to start enjoying your climate sweet spot...and help others to find theirs?

Invite someone to have a Climate Action Conversation with you. Take actions from it. You'll start turning the wheel of Awareness-Actions-Advocacy. You will be initiating what's needed for us collectively to solve climate change now.

*Try the Quick Start Guide*

**Figure 19. Turn the Wheel for Climate Health**

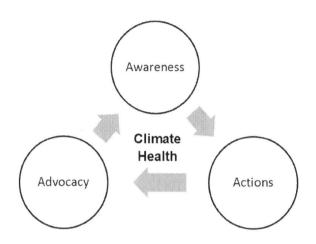

You will be a valuable catalyst starting a chain reaction with each person who becomes involved. Together, we can help people find and enjoy their climate sweet spots. As we do, we will hit the tipping point to support a healthy planet.

*"One good conversation can shift
the direction of change forever."*
~ Linda Lam

Enjoy what you learn and stimulate in your discussions. Please share your experiences from your Climate Action Conversations. I'm eager to support your success.

# Notes

## Chapter 1: We Can Solve Climate Change Now

1. See Ray, Michael L. and Myers, Rochelle, *Creativity in Business: Based on the Famed Stanford University Course That Has Revolutionized the Art of Success* (Broadway, 1986).

2. Buechner, Frederick, *Wishful Thinking: a seeker's ABC* (Harper One, 1993)

3. The Federal Highway Administration statistics report more than 200 million licensed drivers and indicate a typical trip length of approximately 30 miles. [www.fhwa.dot.gov/policyinformation/pubs/hf/pl11028/chapter4.cfm] Greenhouse gas emissions from regular gasoline are 202 grams per mile in tailpipe emissions, and 50 grams per mile in upstream GHG emissions related to the production and distribution of the fuel used to power the vehicle. [Office of Energy Efficiency & Renewable Energy, U.S. Department of Energy, www.fueleconomy.gov/feg/climate.shtml]

4. See Attari, S.Z., Krantz, D.H. & Weber, E.U. Climate change communicators' carbon footprints affect their audience's policy support. *Climatic Change* 154, 529–545 (2019). www.doi.org/10.1007/s10584-019-02463-0

5. Bhowmik, Avit K *et al* Powers of 10: seeking 'sweet spots' for rapid climate and sustainability actions between individual and global scales. *Environ. Res. Lett.* 15 094011, 18 August 2020

## Chapter 2: Shift from Fear to Hope – The Mental Path

1. See Yale Program on Climate Change Communication and George Mason University Center for Climate Change Communication, "Climate Change in the American Mind," December 2020 [www.climatecommunication.yale.edu/publications/climate-change-in-the-american-mind-december-2020/ ]

**2.** See Maruska, Don, *How Great Decisions Get Made: 10 Easy Steps for Reaching Agreement on Even the Toughest Issues* (American Management Association, 2004) and Don Maruska and Jay Perry, *Take Charge of Your Talent: Three Keys to Thriving in Your Career, Organization, and Life* (Berrett-Koehler, 2013).

3. See Whitcomb, Isobel, "Climate anxiety and PTSD are on the rise. Therapists don't always know how to cope." *The Guardian,* April 20, 2021

## Chapter 3: Choose Sufficiency as Your Abundance – The Freedom Path

1. Global Footprint Network (2019) Public Data Package calculates the Biocapacity Deficit for the U.S. at 4.7. See https://data.footprintnetwork.org/#/. Photos are from https://climate.nasa.gov.

## Chapter 4: Focus on Actions Rather than Words – The Change Path

1. Hyde, Catherine Ryan, *Pay It Forward – A Novel* (Simon & Schuster, 1999).

2. For more discussion and examples about how to make effective requests of others, please see pages 100-103 of Maruska, Don and Perry, Jay *Take Charge of Your Talent: Three Keys to Thriving in Your Career, Organization, and Life* (Berrett-Koehler, 2013).

## Chapter 5: Identify What You Love – Your Fun Factor

1. Sharper, Julie, "Lighten Up – According to Science, It's Good for You," *Johns Hopkins Magazine*, Summer 2016

2. See CCL's Volunteer Inventory Form at https://community.citizensclimate.org/resources/item/19/434

3. Tamara Staton references the work of Brown, Stuart, *Play: How It Shapes the Brain, Opens the Imagination, and Invigorates the Soul* (Penguin, 2009), see especially pages 129-141.

## Chapter 6: Promote Awareness – Track Your Carbon Footprints

1. See www.BrightAction.app

2. Hertwich, Edgar G. and Peters, Glen P. "Carbon Footprint of Nations: A Global, Trade-Linked Analysis," *Environ. Sci. Technol.* 2009, 43, 16, 6414–6420 cited in Ghislain Dubois, Benjamin Sovacool, Carlo Aall, Maria Nilsson, Carine Barbier, Alina Herrmann, Sébastien Bruyère, Camilla Andersson, Bore Skold, Franck Nadaud, Florian Dorner, Karen Richardsen Moberg, Jean Paul Ceron, Helen Fischer, Dorothee Amelung, Marta Baltruszewicz, Jeremy Fischer, Françoise Benevise, Valérie R. Louis, Rainer Sauerborn, "It starts at home? Climate policies targeting household consumption and behavioral decisions are key to low-carbon futures," *Energy Research & Social Science*, Volume 52, 2019.

3. Center for Sustainable Systems, University of Michigan. 2021. "Carbon Footprint Factsheet." Pub. No. CSS09-05. September 2021.

Notes

## Chapter 7: Identify Direct Actions – Live More Richly

1. See discussion of the 100 Resource Challenge on pages 88-91 of Maruska, Don and Perry, Jay, *Take Charge of Your Talent: Three Keys to Thriving in Your Career, Organization, and Life* (Berrett-Koehler, 2013).

2. See, for example, Green-e Certified Carbon Offsets (www.green-e.org/certified-resources/carbon-offsets) and efforts to protect ancient and endangered forests, at Canopy Planet (www.CanopyPlanet.org).

3. Poore, J. and Nemecek, T., "Reducing food's environmental impacts through producers and consumers," *Science*, June 1, 2018, pp. 987-992 cited by *CO2 Everything* (www.co2everything.com/co2e-of/oat-milk.

4. Instructions for how to build a simple bioreactor are available online from the Center for Regenerative Agriculture and Resilient Systems at California State University Chico, https://www.csuchico.edu/regenerativeagriculture/bioreactor/bioreactor-instructions.shtml.

5. See Maruska, Don, *How Great Decisions Get Made: 10 Easy Steps for Reaching Agreement on Even the Toughest Issues* (American Management Association, 2004).

## Chapter 8: Find Advocacy opportunities – Share Your Story

1. Noonan, David, "The 25% Revolution—How Big Does a Minority Have to Be to Reshape Society?" *Scientific American*, June 8, 2018.

2. See research by Chenoweth, Erica and Stephan, Maria J., *Why Civil Resistance Works: The Strategic Logic of Nonviolent Conflict* (Columbia University, 2011).

3. See the video or transcript for Session 5, "How Are We Called to Be Public Advocates for Climate Health?" at www.stbenslososos.org/a-beginners-guide-to-creation-care-and-the-climate-crisis/.

4. Bullitt-Jonas, Margaret, correspondence with author.

5. Mission Statement, THIS Is What We Did, www.ThisIsWhatWeDid.org/mission-statement/

## Chapter 9: Select Needs That Interest You – Apply Your Needs Filter

1. Antal, Jim, *Climate Church, Climate World: How People of Faith Must Work for Change* (Rowman & Littlefield, 2018). You can access a free, downloadable study guide for the book, at www.stbenslososos.org/climate-church-climate-world-a-study-guide/.

*Notes*

2. See "A Beginner's Guide to Creation Care and the Climate Crisis," a five-session program with free study guides, video clips, handouts, and scripts. www.stbenslososos.org/climate-church-climate-world-a-study-guide/

**Chapter 10: Create Your Action Portfolio – Celebrate and Share Your Success**

1. Interview with author in February 2022.

2. For further discussion of the Opportunity Sorter tool, see Maruska, Don, *Focus, Freedom, and Fulfillment* (Book IV in the ebook series, *Grow and Enjoy Your Business: Proven Tools for Success*, 2020), Chapter 2, available for free download at www.DonMaruska.com.

# Acknowledgments

Many people and organizations have shaped this book. I will begin with the people who kindly shared their stories. I've organized them in the order in which they appear in the book, and have included some background about them.

Tamara Staton is the Greater Pacific Northwest Regional Coordinator and the Education and Resilience Coordinator for the Citizens' Climate Lobby (CCL)—a nonprofit, grassroots advocacy organization focused on national policies to address climate change. In addition to her CCL endeavors, Tamara is a consultant, facilitator, coach, and educator with over 25 years' experience. She has worked with individuals and groups around leadership development, communication, team building, and improving overall effectiveness. She's passionate about integrating more joy into the climate movement and supporting leaders and climate advocates in building resilience and effective pathways to positive global change. In her downtime, you might find her backpacking, river-running, camping with her family, or strumming her banjo by the creek. She provided in-depth and helpful suggestions to improve this book.

**https://citizensclimatelobby.org**

Calla Rose Ostrander, co-founder of Nerds for Earth, is a catalyst for creating climate healthy communities. She spans awareness, actions, and advocacy in her work throughout the western United States. Calla Rose inspires others with her knack for connecting people who want to help with climate needs.

https://nerds4earth.com

Brian Metcalf, Community Services Director for Rotary de Tolosa Club in San Luis Obispo, California, enjoyed a 38-year career as the founder/owner of an upscale kitchen, bath, and hardware showroom. He is a lifetime avid lover of all things related to nature, and particularly enjoys surfing with his family and friends. A major focus now is serving his community and taking action on climate change through the Rotary Club of San Luis Obispo de Tolosa. Rotary Climate Action Team members live out their motto of "people of action," and have helped me see ways individual efforts can scale to have larger impacts.

https://portal.clubrunner.ca/2851

Jim Thompson has a gift for conceiving and implementing large-scale programs. He founded the Positive Coaching Alliance, which has partnered with roughly 3,500 schools and youth sports organizations nationwide to deliver more than 20,000 live group workshops, reaching over 20 million youth. He also co-founded Recovery Cafe San Jose. He's a Silicon Valley Climate Action Now! member (SV-CAN!), trained as a Climate Reality Project Leader, and is now relentlessly focused on

## Acknowledgments

answering the ultimate question: "What did you do to fight the climate crisis?" with "THIS Is What We Did!" He's a long-time and treasured friend.

**https://ThisIsWhatWeDid.org**

Lisa Altieri, CEO of BrightAction, demonstrates the power of information to help companies, organizations, and cities empower their communities on climate change solutions. She's combined 15+ years of expertise in data science and technical skills with volunteer and community engagement experience. She and her team discover which sources offer the biggest opportunities, and connect people with solutions. I'm grateful for the opportunities to use the collaborative platform, offering simple, everyday actions to make an impact.

**https://BrightAction.com**

Kelly Isabelle DeMarco illustrates a diverse portfolio of personal and professional initiatives for a healthy planet. She wears the hats of occupational therapist, healthcare and sustainability leader, board certified health and wellness coach, and volunteer member of Wisconsin Healthcare Professionals for Climate Action. Her mission is thriving people on a thriving planet! She is dedicated to helping women align lifestyle, purpose, and passion with climate action, and to making the sustainable doable. https://kellygreencoaching.com

The Earthcare Team of St. Benedict's Episcopal Church in Los Osos, California, has nurtured creation care for years. With the leadership and support of The Rev. Caroline Hall and the active engagement of Mike Eggleston, John Horsley, Liz Maruska, Bob Pelfrey, Barry Turner and more, we have developed a whole series of programs. These include study guides for Jim Antal's *Climate Church, Climate World* and a multi-media five-session series "Beginner's Guide to Creation Care and the Climate Crisis" in use by multiple denominations across the United States. With the initiative of the Landscaping Committee and the Parish Council, the congregation has demonstrated how faith-based groups can make a difference locally and beyond.

**www.stbenslososos.org/creation-care/**


The Rev. Dr. Margaret Bullitt-Jonas works in the two Episcopal dioceses in Massachusetts, and the Southern New England Conference, United Church of Christ, to inspire faith-based climate action. Her most recent book (co-edited) is *Rooted and Rising: Voices of Courage in a Time of Climate Crisis.* She's a powerful, personal advocate for climate health, and has even put her body on the line to oppose actions that degrade the planet.

**https://RevivingCreation.org**


Nicole Rycroft is founder and executive director of Canopy Planet, whose mission is to protect the world's forests, species,

## Acknowledgments

and climate, and to help advance Indigenous communities' rights. Her philosophy to "ask for what you want, you might just get it," is foundational to her work in guiding Canopy's team to transform unsustainable supply chains for fast fashion, and advance forest conservation and community rights. She personifies a hopeful, can-do spirit and has earned her recognition as a member of the UBS Global Visionaries Program, Ashoka Fellow, and is a recipient of both the Canadian Environment Award Gold Medal, the Meritorious Service Cross of Canada, and a 2020 Climate Breakthrough Award.

**https://canopyplanet.org**

Seth Bush serves as a coach and trainer for social change leaders and activists. He is a Professional Certified Coach with expertise in training and organizational issues. I met Seth through the Climate Coaching Alliance (https://climatecoaching alliance.org), a volunteer organization mobilizing professional coaches around the world to support a regenerative future. He co-founded the Radical Support Collective, serving leaders for climate action and social justice.

**https://radicalsupport.org/Seth**

The San Luis Obispo Climate Coalition, ECOSLO, Interfaith Power and Light, Episcopal Public Policy Network, and many others have illumined the opportunities to make a difference at home and across the continent.

In addition to those whose stories I've shared in the book, many others contributed to this book. First and foremost, my wife and daughter who share the journey to care for creation. They are my daily renewable energy sources.

John Steinhart, who recruited me to earn my MBA and JD program at Stanford as Director of the Public Management Program, has been a friend and colleague for decades. His many talents include an extraordinary gift for listening and getting to the gist of an issue. John's encouragement and support for this project and many helpful personal connections have shaped me, my work, and this book in important ways.

Michael Ray, professor emeritus of Stanford Business School, and author of *Creativity in Business* and the *Highest Goal*, is a huge inspiration in my career. One of the "live withs" from his creativity course was "do what you love and love what you do." Since those early days, we have remained friends and share interests with the talented members of the Ecological Action Special Interest Group of the Conscious Leadership Guild. I especially appreciate Michael's reviews of multiple drafts, and his patient, thoughtful, and very helpful suggestions.

Art and Judy Stevens have highlighted the power of hope to bring forward our God-given talents and find opportunities to live joyfully together. Their message and life examples encourage my faith that we can find grace-filled solutions to the many challenges we face.

## Acknowledgments

The Conversation Among Masters, a group of master coaches from around the world, encouraged me to write this book. I'm grateful that Raymond Aaron and his team at 10-10-10 Publishing eased the path to publication.

Many others have served as inspirations and teachers in my life. I give particular thanks to my mother, father, and brother. A nurturing, loving family seems like such a rare gift in today's world. How fortunate I have been to have them and all of the people and opportunities I have enjoyed in my life. Thank you.

## About the Author

Don Maruska delights in mobilizing people across the U.S. and world to take climate action and have fun doing it. Since 2003, he has brought together people from scientific, environmental, business, government, and community organizations to boost awareness, actions, and advocacy for stewardship of natural resources. They've developed action plans and concrete results serving the Central Coast of California and communities across the western U.S. From his experience, he's published how to engage people in win-win strategies for ecosystem-based management. Don also served as a founding member of the Ecosystem Advisory panel for NOAA's Pacific Fisheries Management Council, and received an Outstanding Service Award for more than a decade of pioneering leadership.

Don enjoys taking grassroots action by leading the Climate Action Team at his local Rotary de Tolosa Club in San Luis Obispo, California. There he's developed a strategy and actions to become a net zero carbon club. He's also a contributor to the Earthcare Team at St. Benedict's Episcopal Church in Los Osos, California, creating videos, study guides, and support resources. These include the five-part "Beginner's Guide to Creation Care and the Climate Crisis," in use by faith-based groups of multiple denominations across the U.S.

For more than two decades as an author and Master Certified Coach, Don has focused on how to bring forth the best in himself and others. His exploration of the neuroscience and psychology underpinning top performance have confirmed that a hopeful, positive frame of mind gives us our best thinking and capacity to accomplish extraordinary results together. He's the author of *How Great Decisions Get Made—10 Easy Steps for Reaching Agreement on Even the Toughest Issues* (American Management Association, 2004). This work has guided even entrenched adversaries to find productive pathways forward in businesses, families, communities, and government agencies.

With success in coaching people to make the most of their talent and have fun doing it, Don wrote *Take Charge of Your Talent—Three Keys to Thriving in Your Career, Organization, and Life* with co-author Jay Perry (Berrett-Koehler, 2013). This work served as the foundation for creating and directing a nationwide coaching program on behalf of the International City/County Management Association that attracted and served over 10,000 leaders in local government.

As a gift to business and social entrepreneurs during COVID, Don wrote the four-part ebook series, *Grow and Enjoy Your Business—Proven Tools for Success*. The books are available for free download at www.DonMaruska.com.

Don's most successful business and personal ventures have been fun. Whether it was the excitement of launching companies in Silicon Valley to apply new technologies for improved

## About the Author

healthcare, or meeting his wife in a Paris discotheque, each has been a fun experience. That doesn't mean that they have been without challenges. Rather, coming from what he loves to do, he has been able to both enjoy the good times and persevere through the challenges. The results have garnered him a National Innovator's Award and, most importantly, the joy of a loving wife and daughter.

As the first in his family to attend college, Don earned his BA magna cum laude in government from Harvard University. He also completed an MBA and JD from Stanford University, with a Certificate in Public Management.

Don lives on the shores of Morro Bay, California. Nature's powerful presence reminds him of the vital need to restore balance with the essence of what nurtures us.

Visit www.DonMaruska.com for more information about Don's books, speaking, and workshops to bring forth the best in you.

Don is eager to help you put the practices in this book into use and enjoy the results for yourself and the planet. He encourages you to tap the bonus resources found at www.SolveClimateChangeNow.com and to connect with him at Climate@DonMaruska.com.

Made in the USA
Columbia, SC
07 June 2022